Swelling Elastomers in Petroleum Drilling and Development - Applications, Performance Analysis, and Material Modeling

Authored by Sayyad Zahid Qamar, Maaz Akhtar and Tasneem Pervez

Published in London, United Kingdom

IntechOpen

Supporting open minds since 2005

Swelling Elastomers in Petroleum Drilling and Development - Applications, Performance Analysis, and Material Modeling
http://dx.doi.org/10.5772/intechopen.90221
Authored by Sayyad Zahid Qamar, Maaz Akhtar and Tasneem Pervez

Contributors
Sayyad Zahid Qamar, Maaz Akhtar, Tasneem Pervez

First published in London, United Kingdom, 2021 by IntechOpen
IntechOpen is the global imprint of INTECHOPEN LIMITED, registered in England and Wales, registration number: 11086078, 5 Princes Gate Court, London, SW7 2QJ, United Kingdom
Printed in Croatia

British Library Cataloguing-in-Publication Data
A catalogue record for this book is available from the British Library

Additional hard and PDF copies can be obtained from orders@intechopen.com

Swelling Elastomers in Petroleum Drilling and Development - Applications, Performance Analysis, and Material Modeling
Authored by Sayyad Zahid Qamar, Maaz Akhtar and Tasneem Pervez
p. cm.
Print ISBN 978-1-78984-367-5
Online ISBN 978-1-78984-368-2
eBook (PDF) ISBN 978-1-83962-810-8

Meet the authors

Dr. Zahid Qamar, Sayyad is currently working as a professor at the Mechanical and Industrial Engineering Department, Sultan Qaboos University (SQU), Muscat, Oman. He has over 25 years of academic and research experience from different international universities. He has also worked as a professional mechanical engineer in the field for over six years in the heavy engineering and fabrication industry in the following positions: Manager Research and Development; Deputy Manager Design; Production Engineer; and Quality Control Engineer. His research areas are applied materials and manufacturing; applied mechanics and design; reliability engineering; and engineering education. As part of the Applied Mechanics and Advanced Materials Research Group (AM2R) at SQU, he has been involved in different applied research–funded projects in excess of four million US dollars. He has over 200 research/technical publications to his credit (research monographs, edited book volumes, book chapters, publications in refereed international journals and conferences, and technical reports). He has served as associate editor, guest editor, and member of the editorial board for different research journals (including Materials and Manufacturing Processes, Journal of Elastomers and Plastics, the Journal of Engineering Research, American Journal of Mechanical and Industrial Engineering, etc).

Dr. Maaz Akhtar did his bachelor's (Mechanical) and master's (Manufacturing) degrees from NED University of Engineering & Technology (NEDUET), Karachi, and his Ph.D. (Mechanical) from Sultan Qaboos University, Muscat. He spent a few years in the industry and later joined NEDUET in 2005. He is currently serving as an associate professor in the Mechanical Engineering Department at NEDUET. His main research area is the mechanics of materials and manufacturing. He is the author of various research articles published in reputable international journals. He has also presented his work at many international conferences. Most of his recent research work is focused on the use of rubbers, elastomers, and other polymers in different engineering fields.

Prof. Tasneem is currently working as a professor in the Department of Mechanical and Industrial Engineering, College of Engineering at Sultan Qaboos University (SQU). He received a Ph.D. degree in Mechanical Engineering from the University of Minnesota, USA, in 1991. He served as the head of the Department and assistant dean for Postgraduate Studies and Research at SQU, as well as at other universities. Prof. Tasneem is actively engaged in research and consulting work with various national and international organizations. His current research interests are in solid expandable tubulars, swelling elastomers, composite materials, computational mechanics, computer-aided engineering analysis, and engineering education. Apart from teaching, research, and consultancy, Prof. Tasneem has also actively participated in developing new engineering programs and accreditation of different engineering programs by ABET, a US accreditation agency.

Contents

Preface

Swelling elastomers are novel polymers that swell (increase in volume) when in contact with fluids such as water, oil, or acid. They are being increasingly used in many fields ranging from bioengineering to the oil and gas industry. In particular, swelling elastomers are used as sealing elements in many applications in the petroleum industry. Pre- and post-welling material characterization and performance analysis under varying field conditions are very important before their actual deployment. Currently, the placement of such seals is mostly based on a few laboratory tests and a trial-and-error approach. It is very difficult to test all the parameters under all field conditions experimentally. Analytical or numerical techniques are therefore needed to predict the behavior of such elastomers and seals. No comprehensive investigation of this nature is currently available. The main theme of this book is the performance analysis of swelling elastomer seals, using all three investigation methods: experimental, numerical, and analytical. This book consists of various integrated but independent studies.

Currently, there is no published book covering this material: applications of swelling elastomers in the field of petroleum drilling and development; material characterization and performance analysis of these elastomers and seals; and analytical, numerical, and material modeling of these elastomers and applications. The main theme of this book is the analytical, numerical, and experimental investigation of swelling elastomer seals used in petroleum drilling and development applications. This is a research monograph rather than a typical textbook. Its major contributions and applications include insight into the behavior of swelling elastomers and an understanding of swelling phenomenon; performance analysis and optimal selection of swelling elastomers for a given set of field conditions; and design improvement of swelling elastomer packers and other sealing applications.

The book is divided into the following 11 chapters: Chapter 1, "Introductory Chapter: Swelling Elastomers in Petroleum Drilling and Development"; Chapter 2, "Swelling Elastomer Applications in Petroleum Drilling and Development"; Chapter 3, "Swelling Behavior of Elastomers under Water, Oil, and Acid"; Chapter 4, "Experimental Setup for Swellable Elastomers in Cased and Open Holes"; Chapter 5, "Long-Term Integrity Testing of Water-Swelling and Oil-Swelling Packers"; Chapter 6, "Swelling Elastomers: Comparison of Material Models"; Chapter 7, "Mechanical and Structural Behavior of Swelling Elastomers under Compressive Loading"; Chapter 8, "Numerical Investigation of Elastomer Seal Performance"; Chapter 9, "Swelling Elastomers and Tubular Expansion—Numerical Investigation"; Chapter 10, "Analytical Model for Seal Contact Pressure"; and Chapter 11, "New Analytical Model for Swellable Materials".

Some of the major features of this book that make it unique include the following: Experiments are conducted to determine mechanical, structural, and chemical properties before and after swelling for different elastomers exposed to brine solutions of various concentrations. Experimental results are used for performance analysis, as an input for analytical and numerical modeling, and for validation of predicted results. Comparative evaluation of hyperelastic material models has been conducted to identify the best available model that can be used for the simulation of swelling elastomers. The effect of swelling on the mechanical and structural properties of different elastomers has been investigated, both experimentally and numerically, using the best available material model. Performance evaluation of elastomer seals (such as determination of seal contact pressure) has been carried out using the current best material model. As none of the existing models captures the real behavior of swelling elastomers, a new (analytical) material model has been developed. Apart from the hyperelastic nature of such materials, this new model also takes into account the thermodynamics of mixing, diffusion effects, and changes in network entropy. Compared to numerical solutions based on current hyperelastic models, predictions from the new model are significantly closer to experimental results.

The primary audience of this book consists of universities with Masters and Ph.D. programs; petroleum drilling and development organizations; and research organizations, especially in the area of rubbers and polymers, and petroleum drilling/development applications.

The authors gratefully acknowledge the support of Sultan Qaboos University, Petroleum Development Oman, and various other petroleum and rubber companies.

Sayyad Zahid Qamar and Tasneem Pervez
Sultan Qaboos University,
Oman

Maaz Akhtar
N.E.D. University of Engineering and Technology,
Pakistan

Chapter 1

Introductory Chapter: Swelling Elastomers in Petroleum Drilling and Development

Sayyad Zahid Qamar, Maaz Akhtar and Tasneem Pervez

A man ceases to be a beginner in any given science and becomes a master in that science when he has learned that he is going to be a beginner all his life.

Robin George Collingwood

1. Introduction

A major goal of the petroleum industry is to maximize oil production while reducing developmental and operational costs. Maintaining the profitability of old wells and exploiting economically inaccessible new reservoirs are some of the main challenges of petroleum industry [1]. Also, zonal isolation and optimization of the hole size with economic production for both conventional and deep water wells are ongoing problems. A few critical issues such as failure of equipment or application strategy, lack of reliability, complexity of deployment, operational issues, and higher energy requirement for initialization, become hurdles in achieving this target [2]. The relatively new swelling elastomer technology offers innovative and economically viable solutions. It has been successfully tried out in a variety of applications due to its simplicity of design, relatively inexpensive production, and ease of installation and initialization [3]. Petroleum exploration and development industry is witnessing a rapid growth in the use of swelling elastomers. These new applications are aimed at improved oil recovery (IOR) and enhanced oil recovery (and EOR) through slimming of well design, reliable zonal isolation, successful water shutoff, etc. [4]. Initially developed as a mitigation strategy for repair of damaged wells, swelling elastomers are now targeting major savings in cost and time through reduction in borehole diameter, reduced casing clearance, and cementless completions. Swelling elastomers are being used as sealing materials in many applications that target continued profitability of old wells, restarting of production from closed wells, and economic production from inaccessible new reservoirs [5].

Study of swelling elastomers is essential in resolving both application and design problems. Design issues involve seal material, seal geometry, seal performance, rate of swelling, and field conditions. Proper choice of material is of great significance in seal design, as it should possess characteristics that will allow sealing for a long period of time without failure. Material should also be strong enough to bear all types of loads, resist thermal conditions, and have good wear resistance against rock formation. Seal material should also be fast-swelling so that the well can go into production as early as possible. Rate of swelling is very important as it directly affects production, quality, and cost [6]. Proper seal geometry should be used so that seal integrity and reliability can be efficiently maintained. Seal geometry includes thickness and length of sealing elements, number of seals in a series, and gap between sealing elements. Environmental conditions such as temperature, pressure, swelling medium (water, oil, or acid), and formation properties are also very important for seal design. **Figure 1** illustrates the various issues that need to be addressed regarding the use of swelling elastomers and swell packers in the oil and gas industry.

Swell packers, also known as swellable element packers or reactive element packers, are constructed of base pipe similar to the completion tubing in oil and gas wells [7]. Schematic of a typical swell packer is shown in **Figure 2**. Manufacturing of these packers is on a custom-build basis to suit a certain well completion strategy. Application-specific elastomer is molded, thermally cured, and glued (vulcanized) to the base pipe. Back-up (anti-extrusion) rings are integrated into the design in certain cases to keep the elastomer element in place [8].

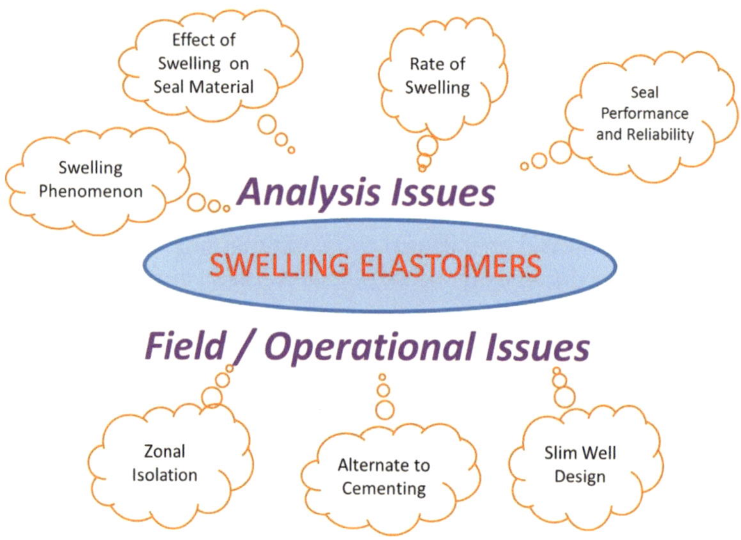

Figure 1.
Various issues related to the use of swelling elastomers and swell packers in the oil and gas industry.

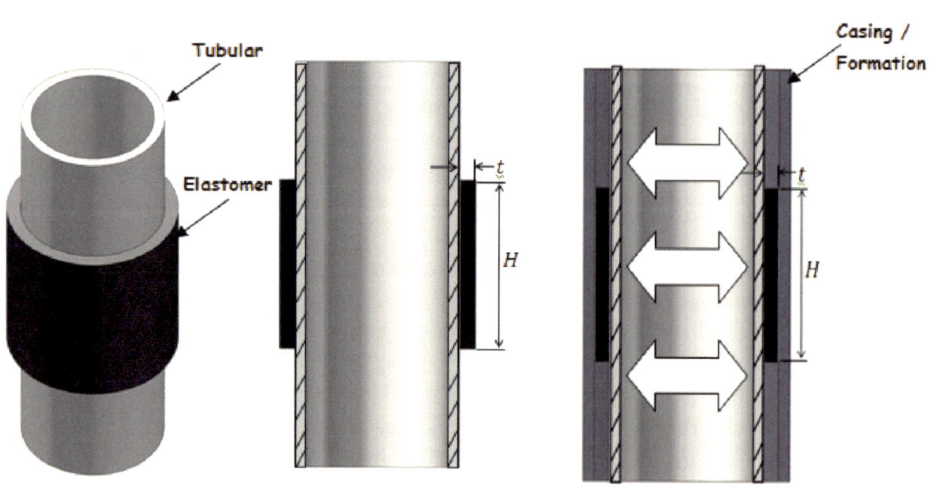

Figure 2.
Schematic of a typical swellable packer.

2. Swelling elastomers

Manufacturing, transportation, communication, mega-structures, virtually each segment of our everyday life is influenced by materials. The development of many technologies that make complex challenging tasks easier has been intimately associated with the accessibility of suitable materials. Advancement in the understanding of a material type is often the predecessor to the stepwise progression of a technology. Metals, ceramics, and naturally occurring polymers have been used as engineering materials for centuries. However, since World War II, the field of materials has been virtually revolutionized by the advent of synthetic polymers.

At room temperature, raw rubbers are amorphous and have linking of only physical nature (entanglements). An agent is required to create a network by producing cross-links. Elastomers are commonly cross-linked by the use of sulfur or peroxide. An elastomer is a type of polymer that has long and flexible molecular structure which has the ability to stretch to several times its original length. It is an ideal material for many applications because it can withstand very large strains. Besides elastic recovery, elastomers have special physical properties (such as flexibility, extensibility, resilience, and durability), which are unmatched by other types of materials.

Swelling elastomer is a new breed of advanced polymer which swells when it interacts with fluids like water, oil, or acid. It is also sometimes referred to as a 'gel' which is an aggregate form of elastomer swollen by immersion into a solvent. Addition of hydrophilic fillers (usually sodium polyacrylate and polyethylene glycol) is needed to make inert elastomers swellable. Examples of some swelling elastomers used in the field are ehtylene propylene rubbers (EPDM), flouroelastomers (AFLAS), hydrogenated nitrile rubbers (HNBR), etc. Karmanova et al. [9] discussed the technology of producing water-swellable elastomers to be used in packer seals for the oil production industry. They used mathematical modeling techniques to predict the effect of hydrophilic additives on the degree of swelling of ethylene-propylene based elastomers.

Swelling results in change in volume, thickness, density, hardness, and other material properties [10–15]. Water-swelling elastomers swell through absorption of saline water through the mechanism of osmosis [16]. To induce swelling in water-based elastomers, large chemical potential gradient is required. Osmotic pressure rises when two solutions of different concentrations are separated by a semi-permeable membrane. When salt dissolved in water becomes mobilized, solute particles collide with the elastomer and transfer momentum, thus generating pressure. This osmotic pressure causes water to flow into the elastomer. Solvent molecules in dilute phase have higher chemical potential than those in concentrated phase; hence this

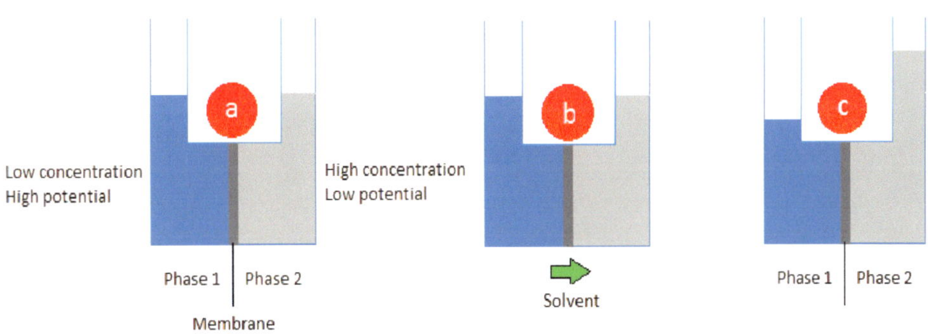

Figure 3.
Fluid influx into the elastomer causes swelling.

concentration gradient (difference in chemical potential) causes solvent flow from solution to elastomer; **Figure 3**.

Swelling in oil-based elastomers takes place by absorption of hydrocarbons through the process of diffusion [16]. Molecules intermingle due to their kinetic energy of random motion. In general, oil-swelling elastomers swell more at higher temperatures and in lighter hydrocarbons. Increase in temperature raises the kinetic energy and decreases the viscosity, resulting in higher swelling rates. For both water-swellable and oil-swellable elastomers, swelling rate depends on temperature, pressure, type of elastomer, and composition of liquid.

3. Brief discussion of applications

Due to the capability of changing properties while interacting with different fluids, swelling elastomers (or gels) are used in a variety of applications in bio-engineering, micro fluidics, and petroleum industry. Biocompatible materials or biosensors that respond to particular molecules (glucose or antigens) are used in drug delivery [17–19]. **Figure 4** schematically shows drug delivery in a bio-responsive gel. Lenses and diapers are other examples of gels [20–22]. Lenses are made of polymers that are hard when dry but readily absorb water and become soft when hydrated. Diapers are superabsorbent materials, absorbing large amounts of liquid but remaining almost dry.

Some researchers and practitioners mathematically and chemically describe a swelling elastomer as a *gel*, because of the rubber consistency after it has absorbed a significant amount of water or other liquid. There are many applications of gels in which volume change takes place. Disposable diapers, human tissues, microfluidic actuators or valves, seals used in different petroleum applications, are some examples. Different biomaterials such as articular cartilage, meniscus, tendons and ligaments, etc. have solid collagen fibrillar network, a fluid phase composed of water, and an ionic phase similar to gels [23]. Soft hydrated tissues swell due to the flow of water and charged ions. Spine consists of intervertebral disc having cartilaginous tissue that connects the vertebral body and allows spine movement. Mechanism of back pain can be understood by studying the mechanics of load transfer in the spine (intervertebral disc), which is affected by swelling of the cartilaginous tissues [24]. One such application, microfluidic valves with fluid–structure interaction, has been described by Zhang et al. [25]. Solid polymeric gel is placed in the fluid flow; volume of the gel starts to increase due to swelling, resulting in closure of the area. **Figure 4** gives a schematic representation of a valve in microfluidics.

Swelling elastomer seals are used in different petroleum applications such as zonal isolation, enhanced oil recovery, water production management, sand

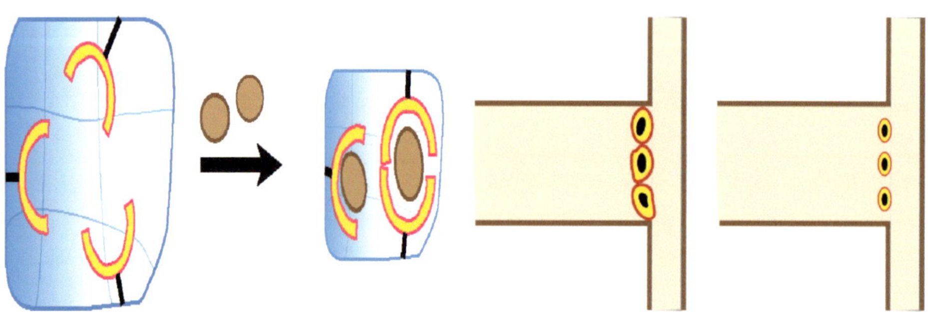

Figure 4.
Drug delivery of bio-responsive gel (left); a microfluidics valve (right).

control, reservoir compartmentalization, production separation, inflow profile control, condensate banking, well completion, slim well technology, intelligent wells, open-hole completions, cased-hole completions, horizontal wells, alternate to cementing, well stimulation, multistage fracturing, etc. More detailed account of the use of swelling elastomers in different petroleum applications is given in Chapter 2.

4. Motivation and significance

Before using a swelling elastomer for any application, its response under different fluids and with different geometries should be taken into account. Erroneous estimation of the elastomer's behavior can result in huge loss of resources and time. Major topic of the current book is performance analysis of swelling elastomer seals and packers for the petroleum drilling and development industry. This assessment can be done through experimental work, numerical simulation, and analytical modeling. All three types of investigations are described in different chapters.

One segment of the current book describes different sets of experiments that will help field engineers choose the correct type and geometry of elastomer seal to be used before deployment in an actual well. Various water-swelling and oil-swelling elastomers are allowed to swell for long periods in salt-water solutions of different salinities and crude oils of different viscosities. Periodic measurements are made for the amount of swelling, density, hardness, tensile properties, compression and bulk properties, etc. Tests are carried out on laboratory scale elastomer samples, and on actual swell packers.

It is sometimes very difficult to test all the parameters experimentally, especially under varying sets of field conditions. On the other hand, a hit-and-trial approach is neither reliable nor efficient, since making a well ready for production needs highly complex, time consuming, and expensive processes. Numerical techniques can be then used to predict the behavior. Appropriate simulations allow us to recreate the physical reality on a computer, and to virtually work out the solutions to these problems. Another section of the book illustrates the use of the finite element method (FEM) to simulate different experiments and seal behavior. This can also help engineers and developers in designing and optimizing the seal geometry for different applications.

Numerical simulation, especially with the proper choice of a material model, can be a powerful tool for performance analysis of swelling elastomers in oil and gas wells. However, a new simulation has to be carried out for every change of rubber or formation material, well type and conditions, etc. This can lead to very high computational costs and time delays. A closed-form solution that predicts the amount of swelling and changes in elastomer properties can be a far more efficient tool for design and development of swelling elastomer applications. One chapter of the book discusses the development of an analytical model for assessment of seal integrity by predicting the seal contact pressure under different conditions and for different seal parameters.

Finite element (FE) analysis requires a suitable material model, a geometrical model of the seal, and proper boundary conditions and applied loads depicting the actual field conditions. Some hyperelastic material models are available for rubber-like materials, and FE simulations using these material models give reasonably accurate results. However, no material model is currently available that mimics the actual behavior of swelling elastomers. FE analysis of higher accuracy is not possible without a more accurate material model. The last chapter of the book takes us through the development of a new material model that can capture the behavior of swelling elastomers more closely, and can be used in future FE simulations.

5. Synopsis of the book

As the title suggests, this book describes applications, performance analysis, and material modeling for swelling elastomers used in petroleum drilling and development. Different sections of the book are dedicated to experimental, numerical, and analytical investigation of swelling elastomer applications in oil and gas wells. Some of the salient features of this research monograph are listed below.

- A thorough literature review has been conducted covering different areas such as applications of inert and swelling elastomer seals and associated technologies, experimental and numerical studies related to swelling elastomers and their applications, material models for hyperelastic and other rubber-like materials, etc.

- Experimental facilities have been designed and constructed, and a variety of experiments have been conducted to understand the effect of various parameters on material properties and behavior of swelling elastomers.

- Three sets of experiments are conducted to determine swelling behavior (change in volume, thickness, hardness, and density), compression properties (stress–strain patterns, elastic modulus, etc), and bulk properties (bulk modulus, Poisson's ratio, etc). These material characterization tests are carried out before swelling and at various stages of swelling. Test parameters (from actual oilfields) include swelling media (water, oil, acid), elastomer materials, water salinities, test temperatures, swelling periods, etc.

- Some of the experimental results are used to determine different structural properties of elastomers such as chain density, and number-average chain molecular weight. Experiments are also conducted to determine the viscosity and density of saline water at particular temperatures. These properties, along with structural properties and specific volume of elastomer, molar volume of liquid, etc. are used to determine the polymer interaction parameter. This is later used in the newly developed material model for parameterization of free swelling.

- Lab tests are conducted on small elastomer samples. No one knows what really happens to sealing elements and swell packers in actual wells. A testing-cum-demonstration unit has been designed and constructed for visualization of the swelling phenomenon, and its effect on swell packers in cased and open holes. Different packers of inert and swelling elastomers are placed in actual petroleum tubulars, and against a concrete wall having sections of different roughnesses, replicating an actual rock formation.

- All tests on swelling elastomers reported in published literature are conducted for a maximum period of a few weeks. However, over long periods of time, even very good elastomers may experience serious material degradation and loss of properties. Durability, integrity, and effectiveness of elastomer seals and packers over long periods is an uncharted area. To address these never-tacked issues, a full-scale longevity test setup has been designed and conducted for actual swell packers of different water-swelling and oil-swelling elastomers, different diametral sizes, maintained at different temperatures over a five-year test period.

- Existing material models that describe the behavior of rubbers and elastomers (hyperelastic materials) are investigated and compared. Compression and bulk experiments are simulated using existing material models and results are compared against empirical data from actual swelling elastomers, before and after different stages of swelling. Material model that gives the closest predictions is selected for further work.

- Numerical modeling and simulation of swelling elastomer seals has been conducted using the best existing material model available in the commercial (nonlinear) finite element package ABAQUS. A large number of simulations are carried out to investigate sealing behavior against different types of elastomer material, field conditions, water salinity, swelling time, seal length, seal thickness, compression ratio, and wall type (rock formation or steel casing). Recommendations are made about the suitability of certain seals for field conditions and parameters in actual oilfields.

- Though numerical simulations can do reasonably accurate performance evaluation, computational time and cost can be large. Mathematical modeling of the behavior of swell packers can be really challenging. However, an analytical model, if developed, can be very time-wise efficient. A closed-form solution has been developed for performance assessment of elastomer packers, especially in terms of the seal contact pressure under different conditions and for different seal parameters.

- A new material model has been developed to closely capture the swelling phenomenon in elastomers, including parameters that were not considered in any of the existing models. This new model is based on the mechanics of swelling,

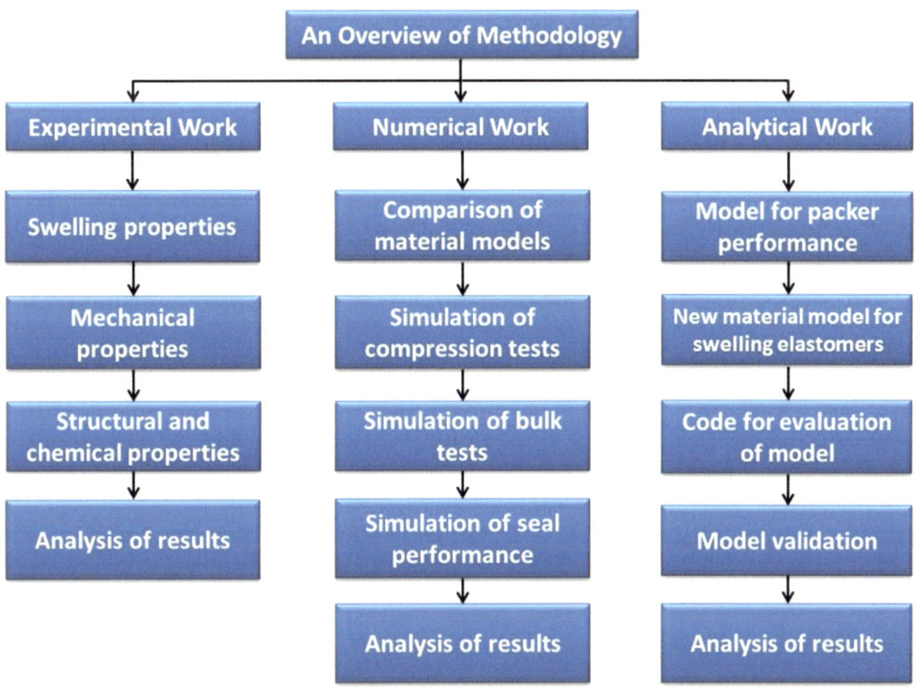

Figure 5.
An overview of methodology for the various studies (experimental, numerical, and analytical) presented in the book.

and it can be used to determine the volume, thickness, etc. after swelling. This model can take actual field conditions as inputs (such as water salinity, operating temperature, mechanical properties, swelling time, etc). Model validation is done using different experimental results.

An overview of methodology for the various modules and studies presented in the book is shown in **Figure 5**.

6. Chapter-wise summary

Chapter-1 introduces the book, and its various sections and chapters. Chapter-2 reviews swelling elastomers and their uses in different petroleum applications. Chapter-3 describes experimental work conducted to investigate the swelling behavior of different elastomers under water, oil, and acid. Chapter-4 describes the design and construction of a demonstration-cum-experimental setup for swell packers in cased and open holes. Chapter-5 discusses longevity testing of water-swelling and oil-swelling packers at different temperatures and pressures, over a five-year period. Chapter-6 presents a comparison of different existing material models (for hyperelastic materials) that can be used to simulate the behavior of swelling elastomers. Chapter-7 explains the determination of different mechanical and structural properties of swelling elastomers under compression, needed for extraction of parameters for numerical simulations. Chapter-8 covers performance evaluation of elastomer seals used in petroleum applications, using FE simulations. Chapter-9 discusses numerical investigation (FEA) of swelling elastomer seals used in conjunction with solid expandable tubulars (SETs). Chapter-10 presents an analytical model for performance evaluation of swell packers in terms of the seal contact pressure. Chapter-11 describes the development of a new material model for swelling elastomers, incorporating mechanics of swelling, thermodynamics of mixing, Gaussian statistics, and other concepts.

Author details

Sayyad Zahid Qamar[1]*, Maaz Akhtar[2] and Tasneem Pervez[1]

1 Mechanical and Industrial Engineering Department, Sultan Qaboos University, Muscat, Oman

2 Mechanical Engineering Department, N.E.D. University of Engineering and Technology, Karachi, Pakistan

*Address all correspondence to: sayyad@squ.edu.om

References

[1] Seyyedattar M, Zendehboudi S, Butt S (2020) "Technical and Non-technical Challenges of Development of Offshore Petroleum Reservoirs: Characterization and Production," Natural Resources Research, 29 (2020), 2147-2189

[2] Sayapov E, Nunez A, Farei I, Benchekor A, Hazzaa O, Mesbahi T, Thomson T, Strom K (2019) "Multistage Frac Zonal Isolation in Extreme HPHT Conditions - Solution to Succeed," Paper # SPE-197653, Abu Dhabi International Petroleum Exhibition & Conference, 11-14 November, Abu Dhabi, UAE

[3] Karmanova OV, Shutilin YF, Moskalev AS, Podvalnyi ES, Petrenko VR, Safonov SV (2018) "Elastomer Seals for Oil Production Equipment," Chemical and Petroleum Engineering, 53 (2018), 642-646

[4] Wang L, Tian Y, Yu X, Wang C, Yao B, Wang S, Winterfeld PH, Wang X, Yang Z, Wang Y, Cui J, Wu Y-S "Advances in Improved/Enhanced Oil Recovery Technologies for Tight and Shale Reservoirs," Fuel, 20 (Dec 2017), 425-445

[5] Mitchell C, Urquhart N (2017) "Development of an Extended Delayed Oil Swell Packer for Deployment in Extended Reach Drilling Oil Wells," Paper # SPE-188912, Abu Dhabi International Petroleum Exhibition & Conference, 13-16 November 2017, Abu Dhabi, UAE

[6] Balasooriya W, Schrittesser B, Pinter G, Schwarz T (2018) "Induced Material Degradation of Elastomers in Harsh Environments," Polymer Testing, 69 (Aug 2018), 07-115

[7] Markovic G, Radovanovic B, Marinovic-Cincovic M, Budinski-Simendic J (2009) "The Effect of Accelerators on Curing Characteristics and Properties of Natural Rubber/Chlorosulphonated Polyethylene Rubber Blend," Materials and Manufacturing Processes, 24(10-11), 10-11

[8] Plavsic M, Pajic-Lijakovic I, Lazic N, Bugarski B, Putanov P (2009) "Catalytic Degradation Processes and Swelling of Alginate Bio-Medical Gels under Influence of Oxygen," Materials and Manufacturing Processes, 24(10-11), 1190-1196

[9] Karmanova OV, Shutilin YF, Moskalev AS, Podvalnyi ES, Petrenko VR, Safonov SV (2018) "Elastomer Seals for Oil Production Equipment," Chemical and Petroleum Engineering, January 2018, 53 (9-10), p 642-646

[10] Ahmed S, Salehi S, Ezeakacha C, Teodoriu C (2019) "Experimental investigation of elastomers in downhole seal elements: Implications for safety," Polymer Testing, 76 (2019), p 350-364

[11] Al-Yami AS, Nasr-El-Din HA, Al-Arfaj MK, Al-Saleh SH, Al-Humaidi AS, Awang MZB, Al-Mohanna KS (2008) "Investigation of Water Swelling Packers," Saudi Aramco Journal of Technology, Fall 2008, p 19-25

[12] Al-Yami A, Nasr-El-Din H, Al-Saleh S, Al-Humaidi A, Al-Arfaj M, Al Moajil A, Awang M, Al-Mohanna K (2008) "Investigation of Oil Swelling Packers," Paper # SPE 113135, SPE Europec/EAGE Conference and Exhibition, Rome, Italy

[13] Al-Yami A, Nasr-El-Din H, Al-Saleh S, Al-Humaidi A, Al-Arfaj M, Awang M, Al-Mohanna K (2008) "Lab Investigation of Oil Swelling Elastomers for Smart Well," Paper # SPE 19403, SPE Offshore Technology Conference, Houston, Texas, USA

[14] Pervez T, Qamar SZ, van de Velden M (2012) “Comparison between Fresh and Exposed Swelling Elastomer,” Journal of Elastomers and Plastics, 44(3), 237-250

[15] Qamar SZ, Al-Hiddabi S, Pervez T, Marketz F (2009) “Mechanical Testing and Characterization of a Swelling Elastomer,” Journal of Elastomers and Plastics, 41(5), 415-431

[16] Herold BH, Marketz F, Froelich BG, Edwards JE, van Kuijk R, Welling RW, Leuranguer C (2006) “Evaluating Expandable Tubular Zonal and Swelling Elastomer Isolation using Wireline Ultrasonic Measurements,” Paper # SPE/IADC 103893, IADC/SPE Asia Pacific Drilling Technology Conference and Exhibition, Thailand

[17] Bromberg LE, Ron ES (1998) “Temperature-Responsive Gels and Thermogelling Polymer Matrices for Protein and Peptide Delivery,” Advanced Drug Delivery Reviews, 31(3), 197-221

[18] Langer R (2000) “Biomaterials in Drug Delivery and Tissue Engineering: One Laboratory's Experience,” Accounts of Chemical Research, 33(2), 94-101

[19] Lee JY, Kang YM, Kim ES, Kang ML, Lee B, Kim JH, Min BH, Park K, Kim MS (2010) “In Vitro and in Vivo Release of Albumin from an Electrostatically Crosslinked in Situ-Forming Gel,” Journal of Materials Chemistry, 20(16), 3265-3271

[20] Qiu Y, Park K (2001) “Environment-Sensitive Hydrogels for Drug Delivery,” Advanced Drug Delivery Reviews, 53(3), 321-339

[21] Carpi F, Frediani G, Turco S, De Rossi D (2011) “Bioinspired Tunable Lens with Muscle-Like Electroactive Elastomers,” Advanced Functional Materials, 21(21), 4152-4158

[22] Suo Z (2012) “Mechanics of Stretchable Electronics and Soft Machines,” MRS Bulletin, 37(03), 218-225

[23] Wilson W, van Donkelaar C, Huyghe JM (2005) “A Comparison between Mechano-Electrochemical and Biphasic Swelling Theories for Soft Hydrated Tissues,” Journal of Biomechanical Engineering, 127(1), 158-165

[24] Schroeder Y, Wilson W, Huyghe JM, Baaijens FP (2006) “Osmoviscoelastic Finite Element Model of the Intervertebral Disc,” European spine journal, 15(3), 361-371

[25] Zhang Y, Liu Z, Swaddiwudhi pong S, Miao H, Ding Z, Yang Z (2012) “Ph-Sensitive Hydrogel for Micro-Fluidic Valve,” Journal of Functional Biomaterials, 3(3), 464-479

Chapter 2

Swelling Elastomer Applications in Petroleum Drilling and Development

Sayyad Zahid Qamar, Maaz Akhtar and Tasneem Pervez

My powers are ordinary. Only my application brings me success.

Sir Isaac Newton.

Abstract

Oil and gas drilling and development is witnessing new and inventive techniques targeted at increased production from difficult and aging wells. As depth of an oil or gas well increases, higher temperatures and harsher environments are encountered. Suitable elastomers can provide good sealing as they possess good resistance to heat and chemical attack, and as they are widely availability at low cost. In comparison with metals, elastomers are lighter in weight and lesser in stiffness and hardness, swell more with increasing temperature, and are usually better in corrosion resistance. Other reasons for their preference include excellent damping and energy absorption, more flexibility and longer life; good sealing even with moisture, heat, and pressure; negligible toxicity; good moldability; and flexible stiffness. As mentioned in chapter-1, swelling elastomers or gels have found extensive use in different applications including drug delivery, microfluidics, biomedical devices, scaffolds for tissue engineering, biosensors, etc. As the main focus of this book is the oil and gas industry, implementation of swelling elastomer technology and deployment in different petroleum applications are discussed below.

Keywords: Swelling elastomers, field applications, oil and gas, drilling and development

1. Introduction

Oil and gas drilling and development is witnessing new and inventive techniques targeted at increased production from difficult and aging wells. As depth of an oil or gas well increases, higher temperatures and harsher environments are encountered. Suitable elastomers can provide good sealing as they possess good resistance to heat and chemical attack, and as they are widely availability at low cost. In comparison with metals, elastomers are lighter in weight and lesser in stiffness and hardness, swell more with increasing temperature, and are usually better in corrosion resistance. Other reasons for their preference include excellent damping and energy absorption, more flexibility and longer life; good sealing even with moisture, heat, and pressure; negligible toxicity; good moldability; and flexible stiffness [1, 2].

As mentioned in chapter-1, swelling elastomers or gels have found extensive use in different applications including drug delivery, microfluidics, biomedical devices, scaffolds for tissue engineering, biosensors, etc. [3–8]. As the main focus of this book is the oil and gas industry, implementation of swelling elastomer technology and deployment in different petroleum applications are discussed below.

2. Swelling elastomer

Elastomers are increasingly being used for sealing and other applications in the oil and gas industry. Specifically developed elastomers possess durable properties and have the ability to withstand detrimental effects of heat, chemicals, and harsh environments. Of special interest is the class known as swelling elastomers. Swelling elastomers are advanced polymers that swell naturally by absorption when exposed to the appropriate swelling agent (oil and/or water). They can be mounted as seals/gaskets directly on to the steel pipe for petroleum piping applications. When in contact with well fluids, they exhibit significant swelling, providing improved sealing against outer tubulars or borehole wall [9, 10].

In conventional drilling, the initial hole has a large diameter, and casing pipe (steel) is run into this hole. Cement is poured in the borehole-casing gap, known as the annulus, to act as a structural support for the well, and to prevent any flow from the formation into the wellbore. Each new pipe should pass through the previous one. Each new casing string thus causes an approximately 20% reduction in the borehole diameter. Any deep well requires multiple casing strings. Because of the "telescoping effect," the well has to begin with a very large-diameter surface casing and finish with a substantially small-diameter production casing. This not only leads to very high costs, but also restricts the exploration and production of oil and gas [11, 12].

Solid expandable tubular (SET) technology has been successfully employed to overcome some of these problems. In SET technology, a conical mandrel is pushed (or pulled) through a solid steel tubular, expanding it to a predetermined size. The most common and valuable application of SET is zonal isolation and water shutoff, preventing fluid cross-flow between geological layers, and reducing the amount of produced water. Together with SET technology, swelling elastomers seals have been used successfully in both open and cased holes. These seals contain several sections of swelling elastomer material mounted on steel pipes or joints. Connecting these joints together can yield seals of any required length. Over time (few days to several weeks), the rubber elements swell considerably, providing a very good sealing. This extremely useful application is also quite low-cost [13]. A crucial brown field was successfully redeveloped in Oman, requiring difficult drilling across fractured carbonate sections, through the use of swelling elastomer seals for zonal isolation [14].

Swellable elastomer is a novel type of unconventional polymer; it increases in volume when exposed to water, oil, or acid. This swelling causes changes in geometry, together with variations in density, hardness and other material properties [15–18]. The mechanism of osmosis is the basis for swelling in water-swell elastomers. On the other hand, volume-change in oil-swell elastomers occurs due to a diffusion action, resulting in the absorption of hydrocarbons into the elastomer material [19, 20]. Rate of swelling is a function of temperature, pressure, elastomer type, and composition of the fluid medium. For applications such as zonal isolation and water shutoff, swelling elastomers are now the inevitable choice. **Figure 1** schematically illustrates the construction of a standard swell packer. There are many applications of swelling elastomers, including

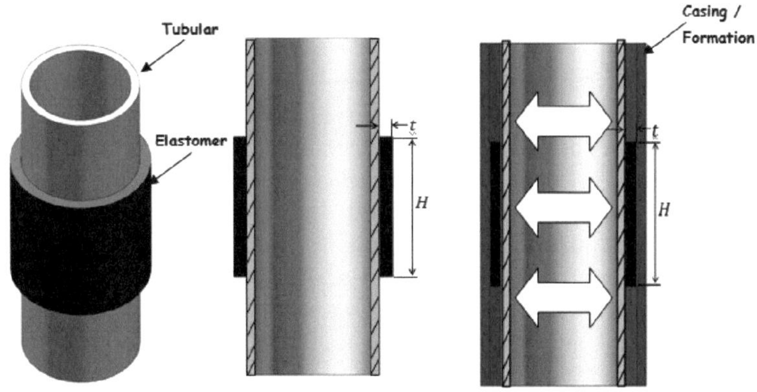

Figure 1.
Arrangement of inner steel tubular, elastomer, and outer casing or rock formation in a typical swellable packer.

profitable production from old wells, restarting of production from discarded wells, and production from difficult-to-access new reservoirs [21, 22].

One of the earliest uses of swelling elastomers in conjunction with an SET system was by Shell in July 2002, known as an open-hole clad (OHC). This was followed by various successful deployments of swell packers in fields requiring zonal isolation in Brunei, Malaysia, Gabon, Nigeria, and the United Kingdom [23]. X-ZIP and E-ZIP solutions (swellable elastomer sections mounted on expandable and standard pipes, respectively) have been success stories in open-hole clads and liners in Oman. Addressing the problems of water shutoff and fracture, these techniques provide new methods for increased oil recovery [24]. Expandable sand screen technology is another good example of the use of swelling elastomers [25, 26]. In recent years, swelling elastomers have become an essential element in many oil well completions since they provide new solutions for zonal isolation and inflow control in horizontal and vertical wells.

3. Applications of swellable elastomers

Many published case studies about swell packers discuss only a few of the related applications. This chapter adopts a more holistic approach, describing most of the key applications of swelling elastomers in the oil and gas industry. For instance, zonal isolation techniques are used to separate undesirable zones from production zones. Safeguarding production lines from water incursion is known as water shutoff [27]. Protection of the wellbore from sand invasion is known as sand control: sand screens provide protection against plugging and wear of well equipment. A well becomes ready to produce through a series of steps collectively known as well completion, cementing being the most critical stage. When existing passages are enlarged or new ones are formed to improve the production the process is called well stimulation [28]. Swelling elastomers have been effectively used in all of these applications, as an integral part of improved oil recovery (IOR) or enhanced oil recovery (EOR), providing low-cost manner and long-term solutions.

Swelling elastomers are generally self-healing, paving the way for easy installation and execution of swell packers, and saving a large amount of rig time and related costs. More relevant case studies about deployment of swell packers and seals in the development of oil and gas fields are presented in the following sections. This brief overview can be beneficial for petroleum students, field practitioners, researchers, and application designers.

3.1 Zonal isolation

Techniques used to stop or reduce the mixing of redundant fluids (from undesirable regions) with production fluids are known as zonal isolation. Conventionally, zonal isolation in a well bore is carried out by cementing the production string in place, and by appropriate use of casing plugs and packers. Swelling elastomer seals used for active zonal isolation are shown in **Figure 2**. Asab field in Abu Dhabi is a mature carbonate reservoir and was drilled in 1985 and horizontally sidetracked in 1999. By the year 2005, water cut increased from 14–25%, considerably reducing the oil production [29]. Water was coming from the fractures due to the failure in the placement of cement plugs at the toe. Deployment of swell packers successfully isolated the unwanted zones, bringing down water cut from 25% to 0.3%. By using swelling elastomers in Malaysia South Furious field, water cut was significantly reduced, and production started within a day, even before complete swelling of the elastomer [23]. Other valuable applications of swelling elastomers for zonal isolation include water production management, reservoir compartmentalization, sand control, inflow profile control, production separation, and control of condensate banking.

3.2 Well completion

As mentioned above, all tasks that are carried out to make the well ready for production are jointly known as *well completion*. These operations include drilling of the hole, running of the casing and cementing, further drilling of the hole until reaching the desired depth, perforation of the steel casing, and stimulation of the cement and formation. Cementing is a critical stage, hydraulically sealing the casing from the formation, thereby isolating individual zones and preventing annular flow. Swellable packers offer cost-effective and more efficient alternatives to traditional cementing and mechanical packer methods. As described above, they consist of polymer segments that swell proportionally when immersed in well fluids. They contain no mechanical parts, and are easy to operate, thereby minimizing cost and time requirements. Swell packers used in well completion are shown in **Figure 3**.

Intelligent well is a unique well type targeting maximum hydrocarbon recovery. Special completion and monitoring tools are installed that can be adjusted either automatically or through human intervention. Use of swelling elastomers in intelligent well completions significantly reduces the development costs and improves the production. Such an optimum zonal isolation cannot be realized in conventional perforating and cementing methods. The offshore Al-Khalij field in Qatar is a success story where corroded and damaged tubulars were remediated by using the intelligent well approach [30]. Using an alternative packer design was tried out before attempting the recompletion. This dependable hydraulic sealing mechanism makes the

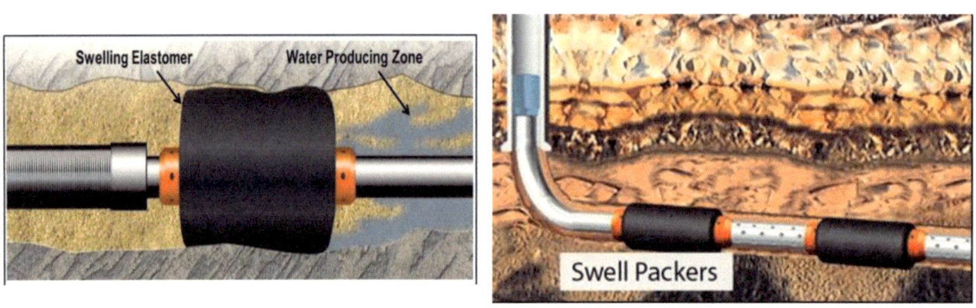

Figure 2.
Schematic of a typical swellable packer used for isolation of water producing zone (left); use of swell packers for zonal isolation in a horizontal well (right).

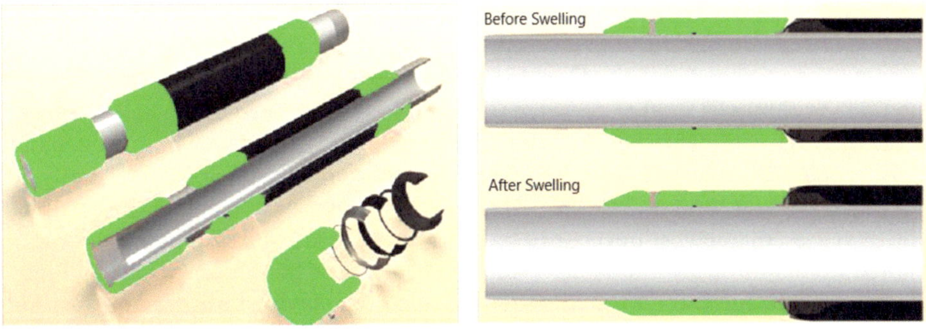

Figure 3.
Elements (and working principle) of a typical swell packer used in well completion operations.

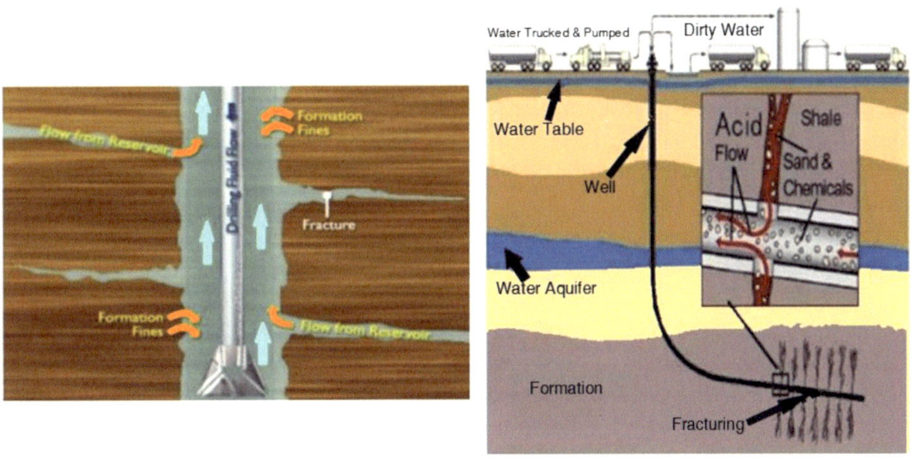

Figure 4.
Two of the more common stimulation operations: Hydraulic fractruring (left), and well acidizing (right).

work-over task technically and economically feasible, especially in the case of mature reservoirs. Use of swell packers guarantees the extension of well life and enhances the oil recovery. Swelling elastomers can also be used as a completion tool in conjunction with open and cased holes, horizontal wells, and SET technology [9, 31].

3.3 Stimulation operations

It is sometimes required to enlarge old channels or create new ones in the production zone of a well. This is accomplished by the methods of acidizing, formation fracturing, etc. and the process is called *stimulation* [32]; **Figure 4**. The stimulation technique of fracturing consists of opening of new flow channels in the rock formation surrounding a producing well. This increases the surface area through which formation fluids can flow into the well, and also extends beyond possible fractures close to the wellbore. The in-situ well conditions can pose challenges in the use of swell packers in stimulation operations. These include local temperature and pressure, and shrinkage forces of the tubular. Moreover, certain thermal effects can result in contraction of the seal: owing to temperature drop because of contact with stimulation fluid. Successful results were reported by an operator in USA in stimulation efforts in horizontal open-hole completions. Technologies used were expandable liner hanger, swelling elastomer packers, and ball drop sleeves [33]. Some other successful applications of swellables in stimulation operations are hydraulic fracturing, multi-stage fracturing, and matrix acidization.

3.4 Underbalanced drilling

It is a conventional practice to go for overbalancing in well drilling. In this method, a column of fluid of preselected density is maintained in the hole to serve as the primary well-control mechanism. By design, pressure at well bottom is kept higher than the formation pressure. During *underbalanced drilling* (UBD) of oil and gas wells, a lighter fluid replaces the fluid column, and the pressure on the bottom of the well is designed on-purpose to be lower than the pressure in the formation; **Figure 5**. This is done to allow formation fluids to rise to the surface while drilling, preventing damage to the formation being drilled. The hydrostatic head of the fluid may naturally be less than the formation pressure, or it can be induced by adding different substances to the liquid phase of the drilling fluid, such as natural gas, nitrogen, or air [34].

One use of underbalanced drilling is to find out probable thief zones for proper placement of swell packers [35]. Effective zonal isolation can be achieved through a judicious combination of UBD and swellable elastomers. This can lead to maximization of well performance and significantly improved hydrocarbon recovery. In the case of underbalanced drilling in Nimr reservoir in Oman [36], it was concluded from data extracted by UBD that rogue fractures and thief zones are important factors in the movement of water. Swell packers were then deployed for water shutoff in zones detected by UBD.

3.5 Enhanced oil recovery

Any technique used to increase the amount of oil that can be recovered from a reservoir is known as enhanced oil recovery (EOR). This is generally achieved by injecting a suitable material into an existing well, increasing the pressure and reducing the oil viscosity. EOR is a thermal or compositional transformation of either the hydrocarbons or reservoir rock to aid in the recovery of additional volumes. EOR helps to maximize the oil reserves recovered, extend the life of fields, and increase the recovery factor. EOR techniques can be used for higher recovery from dwindling old reservoirs or difficult new ones. An openhole packer system is shown in **Figure 6**, for effective zonal isolation in complex offshore deep water wells, unconventional completions, and mature oil field locations.

By combining swellable elastomers and primary cementing, it is possible to successfully isolate oil (or gas) and water. Employing miscible gas injection, the

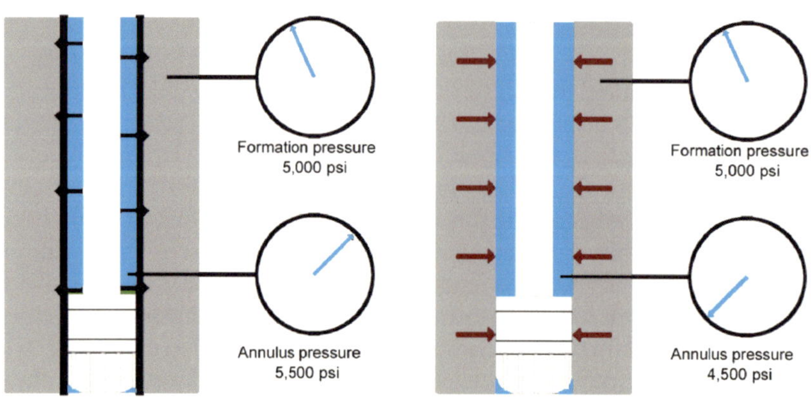

Figure 5.
Schematic illustration of conventional overbalanced drilling (left), and an underbalanced drilled well (right).

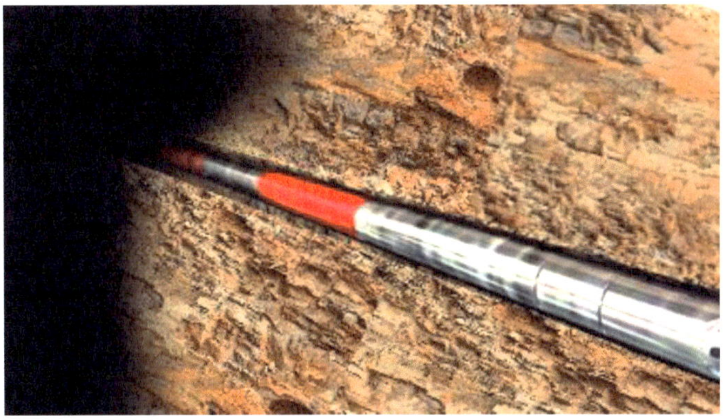

Figure 6.
Openhole packer system for EOR applications requiring maximum isolation efficiency in difficult wells.

first EOR scheme in southern Oman was implemented by Petroleum Development Oman (PDO) in the Harweel Cluster. Efficient zonal isolation was a prerequisite for enhanced hydrocarbon recovery. As cementing was unsuccessful in many cases, swellable elastomers were used and produced very positive results [37]. Durongwattana et al. [38] declare that quite a few EOR based well technologies have matured to a level that they can be considered to be field proven and cost effective. They particularly mention the use of expandable clads and swellable elastomers in applications such as shutoff of high permeable zones and fractures, and well segmentation as a remediation action after breakthrough of water or gas in the segment.

3.6 Evaluation of swellable packers

As swelling elastomer seals are relatively new in the oil and gas industry, reliable performance assessment is critical. One method used to evaluate if swell packers have effectively achieved zonal isolation is wireline ultrasonic measurements. Herold et al. [19] describe conventional and new ultrasonic tools for the evaluation of zonal isolation through swelling elastomers. Perhaps the most promising ultrasonic technique is the third interface echo (TIE) method, ensuring good hydraulic seal and isolation of zones through swellable elastomers [39]. Some other studies have been carried out for performance assessment of swellable packers and seals in petroleum drilling using experimental, analytical, and numerical methods [40–42].

4. Future trends

Already showing major successes, new swelling elastomers and applications are still being developed. As with every new technology, there are many challenges that need to be addressed. Use of swell packers in the extreme conditions of high-pressure and high-temperature (HPHT) wells is uncertain. Reservoirs that are more chemically aggressive (highly acidic, for instance) are also uncharted waters. Controlling the speed and amount of swelling, and activation of swelling in especially targeted settings, are also areas where more work is required. Development of new swelling-elastomer materials and designing of improved and innovative applications is a need of the moment for the working envelope of HPHT and aggressive reservoirs [43, 44].

5. Chapter summary

Swelling packers and seals offer promising unconventional solutions in well completion, repair of damaged wells, water shutoff and zonal isolation, and other downhole applications. Apart from functional performance, swellables are also more cost effective and economical in terms of rig time. As they offer improved production volume and higher rate of hydrocarbon recovery, they are further attractive due to lower environmental impact. In the case of failure of tubular or cementing, or in situations of high water-cut, swelling elastomers have been successfully and efficiently used for well remediation and recompletion. However, the technology has great potential, and should be introduced (if expedient) during the well design phase rather than as only a remediation method. This can save a lot of time and cost spent in firefighting after well damage or shutoff. The review presented above can assist field engineers in the selection of suitable swellable elastomer devices, and can be useful to research and development (R&D) engineers in the design of new downhole applications.

Author details

Sayyad Zahid Qamar[1*], Maaz Akhtar[2] and Tasneem Pervez[1]

1 Mechanical and Industrial Engineering Department, Sultan Qaboos University, Muscat, Oman

2 Mechanical Engineering Department, N.E.D. University of Engineering and Technology, Karachi, Pakistan

*Address all correspondence to: sayyad@squ.edu.om

References

[1] Trelleborg Oil and Gas Sealing Solutions (2019), www.tss.trelleborg.com; accessed May 2019

[2] Patel H, Salehi S, Ahmed R, Teodoriu C (2019) "Review of Elastomer Seal Assemblies in Oil and Gas Wells: Performance Evaluation, Failure Mechanisms, and Gaps in Industry Standards," Journal of Petroleum Science and Engineering, 179 (2019), p 1046-1062

[3] Bromberg LE, Ron ES (1998) "Temperature-Responsive Gels and Thermogelling Polymer Matrices for Protein and Peptide Delivery," Advanced Drug Delivery Reviews, 31(3), 197-221

[4] Buenger D, Topuz F, Groll J (2012) "Hydrogels in Sensing Applications," Progress in Polymer Science, 37(12), 1678-1719

[5] Carpi F, Frediani G, Turco S, De Rossi D (2011) "Bioinspired Tunable Lens with Muscle-like Electroactive Elastomers," Advanced Functional Materials, 21(21), 4152-4158.

[6] Langer R (2000) "Biomaterials in Drug Delivery and Tissue Engineering: One Laboratory's Experience," Accounts of Chemical Research, 33(2), 94-101.

[7] Qiu Y, Park K (2001) "Environment-Sensitive Hydrogels for Drug Delivery," Advanced drug delivery reviews, 53(3), 321-339.

[8] Suo Z (2012) "Mechanics of Stretchable Electronics and Soft Machines," MRS Bulletin, 37(03), 218-225.

[9] Kennedy G, Lawless A,Shaikh K, Alabi T (2005) "The Use of Swell Packers as a Replacement and Alternative to Cementing," Proceedings SPE Annual Technical Conference and Exhibition, 9-12 October 2005, Dallas, Texas, SPE Paper # 95713-MS, Society of Petroleum Engineers

[10] Polgar LM, Fallani F, Cuijpers J, Raffa P, Broekhuis AA, van Duin M, Picchioni F (2017) "Water-swellable elastomers: synthesis, properties and applications," Reviews in Chemical Engineering, 35 (1), p 45-72

[11] Bellarby J (2009), Well Completion Design, Elsevier

[12] Blasingame K, Cales G (2003), "Solid Expandable Tubular Technology in Mature Basins" AAPG Eastern Section Meeting, Pittsburgh, PA, September 6-10, 2003

[13] Hembling D, Salamy S, Qatani A, Carter N, Jacob S (2006) "Swell Packers: Enabling Openhole Intelligent and Multilateral Well Completions for Enhanced Oil Recovery," Proceedings IADC/SPE Asia Pacific Drilling Technology Conference, 13-15 November 2006, Bangkok, Thailand, SPE Paper # 100824-MS

[14] Marketz F, Leuranger C, Welling RWF, Ogoke V (2005) "Waterflood Appraisal — Well Delivery with Expandable Tubulars," Proceedings 2005 International Petroleum Technology Conference, SPE Paper # IPTC 10345, p 507-515

[15] Qamar SZ, Hiddabi SA, Pervez T, Marketz F (2009) "Mechanical Testing and Characterization of a Swelling Elastomer," Journal of Elastomers and Plastics, 41 (5), September 2009, p 415-431

[16] Pervez T, Qamar SZ, Siddiqui RA, van de Velden M (2009) "Effect of Exposure on Material Response of a Swelling Elastomer," Archives of Materials Science and Engineering, 37 (2), June 2009, p 77-84

[17] Pervez T, Qamar SZ, Van de Velden M (2012) "Comparison between Fresh and Exposed Swelling Elastomer,"

Journal of Elastomers and Plastics, 44 (3), 2012, p 237-250

[18] Qamar SZ, Pervez T, Akhtar M, Al-Kharusi MSM (2012) "Design and Manufacture of Swell Packers: Influence of Material Behavior," Journal of Materials and Manufacturing Processes, 27 (7), p 721-726, 2012.

[19] Herold BH, Edwards JE, Kujik RV, Froelich B, Marketz F, Welling RWF, Leuranguer C (2006) "Evaluating Expandable Tubular Zonal and Swelling Elastomer Isolation using Wireline Ultrasonic Measurements," Paper # SPE/IADC 103893, IADC/SPE Asia Pacific Drilling Technology Conference and Exhibition, 13-15 November 2006, Bangkok, Thailand.

[20] Qamar SZ, Akhtar M, Pervez T (2015) "Swelling Behavior of Soft and Biological Materials," Key Engineering Materials, Vol 656-657, 2015, p 580-585

[21] Qamar SZ, Akhtar M, Pervez T, Al-Kharusi MSM (2013) "Mechanical and Structural Behavior of a Swelling Elastomer under Compressive Loading," Materials and Design, Vol 45, March 2013, p 487-496

[22] Pervez T, Seibi AC, Al-Hiddabi SA, Al-Jahwari FK, Qamar SZ, Marketz F (2007) "Solid Tubular Expansion in Horizontal Wells," SPE Paper # 105704, 15th SPE Middle East Oil and Gas Show and Conference, 11-14 March 2007, Manama, Bahrain.

[23] Kleverlaan M, van Noort RH, Jones I, (2005) "Deployment of Swelling Elastomer Packers," Journal of Petroleum Technology, 57 (9), September 2005, p 45-46

[24] Al-Balushi HM, Al-Rashdi MH, Al-Shandoodi MH, van Noort RH (2004) "World's Longest Expandable Open Hole Clad and Open Hole Liner with Swelling Elastomer Deployed in Yibal Horizontal Well", Proceedings IADC/SPE Asia Pacific Drilling Technology Conference and Exhibition, SPE Paper # 88026-MS

[25] Wood P-C, Duhrkopf D, Green A (2007) "Qualifying a New Expandable Reservoir Completion System," Proceedings SPE/IADC Drilling Conference, Volume 2, 2007, SPE Paper # 105556-MS

[26] Abdul-Rahman S, Derek L, Jit JL, Kuo CO (2006) "Innovative use of Expandable Sand Screens Combined with Propped Hydraulic Fracturing Technology in Two Wells with Intelligent Completions in the Egret Field, Brunei: Challenges, Successes, and Best Practices Learned," Proceedings SPE Asia Pacific Oil and Gas Conference and Exhibition 2006, SPE Paper # 101187-MS

[27] Akhtar M, Qamar SZ, Pervez T (2012) "Swelling Elastomer Applications in Oil and Gas Industry," Journal of Trends in the Development of Machinery and Associated Technology, 16 (1), 2012, p 71-74

[28] Akhtar M, Qamar SZ, Pervez T (2018) "Some Downhole Applications of Swelling Elastomers in Oil and Gas Wells," International Journal of Petrochemical Science & Engineering, 2018, 3 (3), p 75-78

[29] Al-Saiid KA, Abdalla E, Cherif M, Elliot SP (2007) "Practical Uses of Swellable Packer Technology to Reduce Water Cut: Case Studies from the Middle East and Other Areas," Paper # SPE 108613-MS, Offshore Europe, 4-7 September 2007, Aberdeen, Scotland, U.K.

[30] Charles PE, Jaafar MR, George BC, Gjelstad G, Elliot SP (2008) "Corroded Casing: Testing of Sealing Capability and Retrievability of a Swellable Elastomer Packer," Paper # SPE 116210-MS, SPE Annual Technical Conference and Exhibition, 21-24 September 2008, Denver, Colorado, USA.

[31] Kalyani T, Ram A, Kumar SA, Raj SR, Badwaik DZ (2009) "Swellable Packers in Unique Horizontal Completions Solves Difficult Challenges in Offshore India Vasai East Field," Paper # SPE 124385-MS, Offshore Europe, 8-11 September 2009, Aberdeen, UK.

[32] Tipton DS (2014) "Mid-Continent Water Management for Stimulation Operations," Paper # IDSPE-168593, SPE Hydraulic Fracturing Technology Conference, 4-6 February 2014, the Woodlands, Texas, USA

[33] Rutger E, Dustin AY, Gregory WV, Kristian S (2009) "Design Methodology for Swellable Elastomer Packers in Fracturing Operations," Paper # OTC-20157-MS, Offshore Technology Conference, 4-7 May 2009, Houston, Texas, USA.

[34] Ojha K, Saxena A, Pathak AK (2014) "Underbalanced Drilling and Its Advancements: An Overview," Journal of Petroleum Engineering and Technology, Vol 4, No 2

[35] John R (2006) "Underbalanced Drilling in the Reservoir; An Integrated Technology Approach," Paper # SPE 103576-MS, SPE Russian Oil and Gas Technical Conference and Exhibition, 3-6 October 2006, Moscow, Russia.

[36] Richard SM, Deepankar B, Suryanarayana PV (2004) "Impact of Thief Zone Identification and Shut-off on Water Production in the Nimr Field," Paper # SPE/IADC 91665, SPE/IADC Underbalanced Technology Conference and Exhibition, 11-12 October 2004, Houston, Texas, USA.

[37] Laws MS, Elliot FJ, Harry S, Neale C (2006) "PDOBs Proactive Approach to Solving a Zonal Isolation Challenge in Harweel HP Wells Using Swell Packers," Paper # SPE 100361-MS, IADC/SPE Asia Pacific Drilling Technology Conference and Exhibition, 13-15 November 2006, Bangkok, Thailand.

[38] Durongwattana N, Toempromraj W, Jedsadawaranon P, Sompopsart S (2011) "Well Integrity Remediation — A Challenge for Swellable Technology," International Petroleum Technology Conference, 15-17 November 2011, Bangkok, Thailand

[39] Al-Kindi ZL, Al-Suwaidi AS, Ibrahim ME, Kamal J, Al Marri MF, Emad S, Antoine E (2009) "Increased Certainty in the Determination of Zonal Isolation Through the Integration of Annulus Geometry Imaging and Improved Solid-Fluid Discrimination," Paper W# SPE 120061-MS, SPE Middle East Oil and Gas Show and Conference, 15-18 March 2009, Bahrain

[40] Akhtar M, Qamar SZ, Pervez T, Al-Jahwari FK (2018) "Performance Analysis of Swelling Elastomer Seals," Petroleum Science and Engineering, 165 (2018), p 127-135

[41] Qamar SZ, Pervez T, Akhtar M (2016) "Performance Evaluation of Water-Swelling and Oil-Swelling Elastomers," Journal of Elastomers and Plastics, 48 (6), 2016, p 535-545

[42] Al-Hiddabi SA, Pervez T, Qamar SZ, Al-Jahwari FK, Marketz F, Al-Houqani S, van de Velden M (2015) "Analytical Model of Elastomer Seal Performance in Oil Wells," Applied Mathematical Modelling, 39 (10-11), June 2015, p 2836-2848

[43] Chen X, Bartos J, Salem H, Zonoz R (2016) "Elastomers for High Pressure Low Temperature HPLT Sealing," Paper # IDOTC-27227, Offshore Technology Conference, 2-5 May 2016, Houston, Texas, USA

[44] Seyger R, Resink S, Harms H, Hibberd R (2013) "The Future of Swelling Elastomers: An Elastomer Manufacturer's View of Swelling Elastomer Developments and Market Trends," the Journal of Engineering Research, Vol. 10, No. 1, 50-64

Chapter 3

Swelling Behavior of Elastomers under Water, Oil, and Acid

Sayyad Zahid Qamar, Maaz Akhtar and Tasneem Pervez

There are three principal means of acquiring knowledge: observation of nature, reflection, and experimentation. Observation collects facts; reflection combines them; experimentation verifies the result of that combination.

Denis Diderot

Abstract

It is very important to determine the behavior of elastomer materials under realistic well conditions in order to select appropriate swelling elastomers for a particular set of field conditions, for successful modeling and simulation of various downhole processes, and for design improvement of swell packers and other sealing applications. In collaboration with national and regional petroleum development and rubber engineering companies, a series of experimental studies were therefore conducted at Sultan Qaboos University for characterization of swelling related material behavior of different elastomers. Results from some of these investigations (studies A, B, and C) are reported and discussed in this chapter.

Keywords: swelling behavior, water-swelling, oil-swelling, acid induction

1. Introduction

It is very important to determine the behavior of elastomer materials under realistic well conditions in order to select appropriate swelling elastomers for a particular set of field conditions, for successful modeling and simulation of various downhole processes, and for design improvement of swell packers and other sealing applications [1]. In collaboration with national and regional petroleum development and rubber engineering companies, a series of experimental studies were therefore conducted at Sultan Qaboos University for characterization of swelling related material behavior of different elastomers. Results from some of these investigations (studies A, B, and C) are reported and discussed in this chapter.

2. Experimental setup

There are standard procedures for conducting most of the tests on rubber materials. It is important to follow these procedures carefully in conducting the tests in order to obtain consistent results. Methodology of the swelling test was developed in consultation with petroleum engineers and rubber manufacturers. Other experiments were designed and performed in line with standard ASTM test methods. Apart from regularly available testing equipment, some simple test rigs and fixtures were designed and fabricated. The tests include swelling behavior (volume, thickness, and hardness change), compression set, tensile set, and tensile properties. Values of test temperature, water salinity, oil viscosity, and acid concentration were selected to emulate actual well conditions in different local oilfields.

2.1 Sample types

Plate samples are used to mimic the actual response of a swell packer (rubber element vulcanized onto a petroleum casing). These generally consist of 6 mm thick elastomer mounted on approximately 50 x 50 x 2.5 mm steel plates. As a basis for comparison, disc samples (28 mm diameter, 12 mm thickness) are used to assess the behavior if the elastomer is allowed to swell freely in all directions. It is very difficult to prepare standard dumbbell shape tensile specimens from elastomer sheet material, as the material is really flexible and not firm. It is far easier to perform tension test using ring samples, in accordance with ASTM standards. All sample elastomer materials are provided by different regional oilfield operators and rubber companies.

The three sample geometries are shown in **Figure 1**. Disc samples represent free swelling, and are used to measure compression set, swelling volume and thickness, hardness, and compression and bulk properties. Plate samples characterize restricted swelling (to replicate actual seal behavior; elastomer mounted on pipe); and are used to measure swelling volume, thickness, and hardness. Ring samples are used to measure tensile set, and tensile properties.

2.2 Sample preparation

Elastomer samples (in finished form) are sometimes supplied directly by rubber manufacturers, but are mostly prepared in-house. If swell packers (elastomer mounted on a base pipe) are provided, pipe is cut into sections on a lathe machine, and pipe sections are cut into desired plate samples using milling or saw cutting machine. To get disc and ring samples, elastomer is removed from the packer, and surface grinding is done on these sheets to smooth out the roughness, and to get the required thickness. If the elastomer is supplied as sheets, square or rectangular pieces of requisite size are cut, and then mounting/vulcanizing on pre-cut steel plates is done using specific glues. Disc and ring samples are cut directly from the sheet. **Figure 2** shows the various steps involved in sample preparation.

Figure 1.
The three sample geometries used for elastomer testing: Disc, plate, and ring.

Figure 2.
Different stages in sample preparation: Cutting sections from packer, and plate samples from sections; removing elastomer from pipe; surface grinding; cutting discs and rings using die-and-punch set.

2.3 Swelling medium

For water-swelling elastomers, salt-water solutions of different concentrations are prepared inhouse. For oil-swelling elastomers, actual crude oil from regional oilfields is procured from petroleum companies. For acid testing, HCl solutions of requisite concentration are prepared inhouse. Total testing time varies from one to three months. Swelling measurements are taken at different intervals, such as on day-zero (before swelling), and after 1, 3, 4, 7, 15, and 31 days of swelling.

2.4 Hardness test

An instrument called a durometer is used to determined hardness of rubbers and elastomers; **Figure 3**. A blunt indenter point is pressed into the sample surface (without causing any puncturing), and the instrument measures the resistance to penetration through the action of a spring. In analog models, movement of a pointer across a scale to indicates the resistance to penetration. The Shore Durometer is scaled from 0 to 100, higher numbers representing higher hardness. The two commonly used scales are “A” for soft rubbers and “D” for harder materials. Applicable test standard is *ASTM D2240* [2]. Room temperature hardness values reported here are average of five readings taken at different locations on the same each sample.

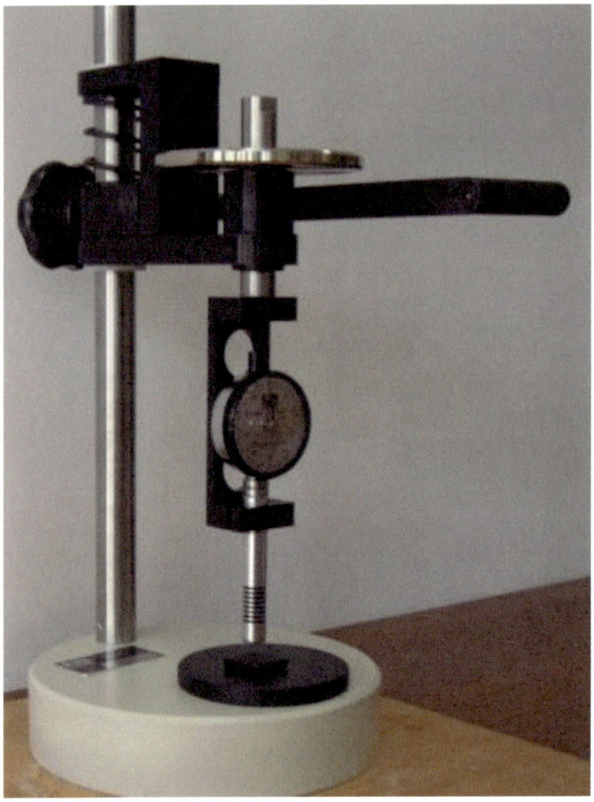

Figure 3.
Durometer for hardness testing.

2.5 Compression set test

Compression set is a measure of the ability of a rubber or elastomer to retain elastic properties after prolonged action of compressive stress. The test is run for either 22 hours or 70 hours. The height that is not recovered represents the compression set, reported as a fraction (percentage) of the amount by which a standard test sample fails to return to its original thickness when acted upon by a standard compressive force/deflection for a predetermined time period at a specific temperature. Following ASTM guidelines (*ASTM D395, method B*) [3], a dedicated fixture was designed and fabricated in-house for the compression set test of disc-type elastomer specimens (13 mm diameter, 6 mm thickness); **Figure 4**. Each test is carried

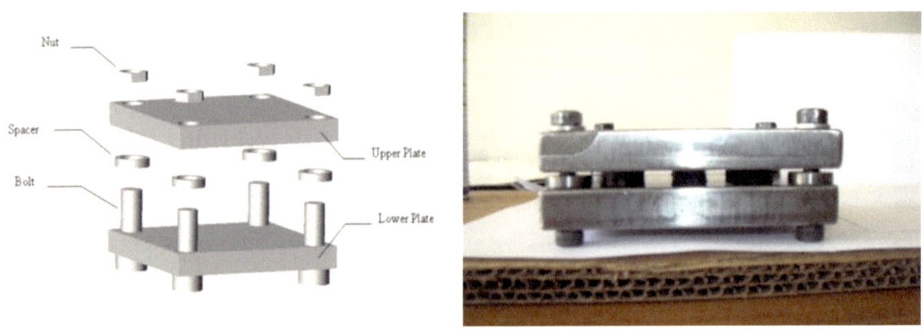

Figure 4.
Test fixture for compression-set test, fabricated in-house.

out at a specified temperature. Samples are then removed to room temperature and allowed to cool for 30 minutes before thickness measurements. Rubbers having god resistance to compression set recover significantly upon releasing the load. 100% recovery is not necessary for an elastomer to work as an effective and repeatable seal. Moreover, if there is a constant compression on the seal, material recovery is not very important.

2.6 Tensile set test

When a specimen is stretched to twice its original size (100% stretch) for a stipulated time, then allowed to recover for the same time at room temperature, the remaining amount of extension determines the tensile set value. A special tensile fixture was designed and fabricated (in line with ASTM guidelines) for this test; **Figure 5**. Applicable standard is *ASTM D412* [4].

2.7 Tensile properties test

This test is used to quantify elastomer behavior under axial tensile loading at room temperature. After the resulting data is plotted as a stress–strain graph, tensile properties can be evaluated; such as modulus of elasticity, tensile strength, and % elongation (or % area reduction). Apparatus used is a universal testing machine fitted with a small load cell for rubbers and elastomeric materials, specially designed and fabricated hook-type grips to hold ring samples (in line with ASTM guidelines), data acquisition and recording system, and ring-type elastomer specimens (16 mm inside diameter and 3 mm thickness); **Figure 6**.

The following test procedure is carried out. Universal testing machine is carefully calibrated. Elastomer ring is mounted into the universal testing machine, using hook-type fixture. As specified in ASTM D412 [4] test standard, machine is set for a speed of 500 ± 50 mm /min. Values of the applied force and consequent elongation are recorded. Average readings from three samples are tabulated, and converted to stress–strain plots. From the data and graphs, tensile properties are calculated.

2.8 Swelling test

The objective of the test is to find the amount of progressive swelling (volume and thickness) and density change in an elastomer for a specified test period, and at different temperatures. Water-swelling elastomers are exposed to salt solutions

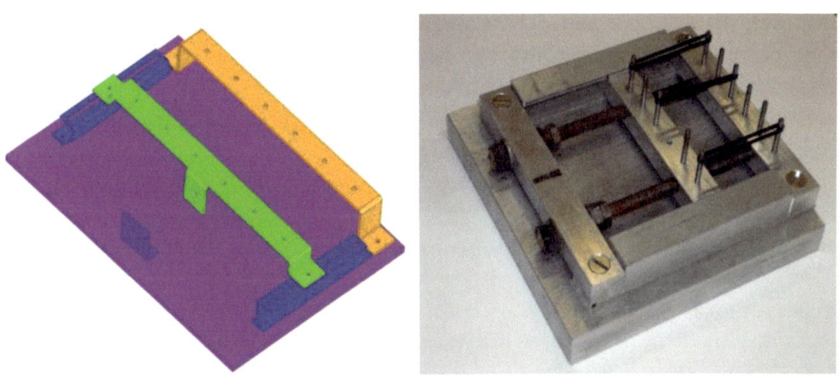

Figure 5.
Test fixture for tensile-set test, fabricated in-house.

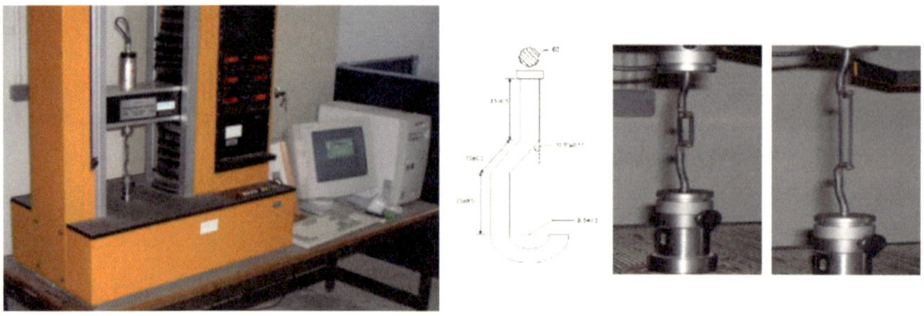

Figure 6.
Apparatus for tensile testing: Universal testing machine with data acquisition system, and hook type fixture for ring samples.

of different concentrations (ranging from 6,000 ppm or 0.6%, to 200,000 ppm or 20%), while oil-swelling elastomers are immersed in crude oils of different viscosities. Disc samples are used to study the swelling response of free (unconfined) elastomer, while plate samples (elastomer vulcanized onto steel plate) are used to replicate the sealing behavior of elastomer mounted on a pipe. Samples are placed in temperature-resistant sealable glass jars containing proper swelling medium (brine, oil, acid), utilized to maintain constant concentration even at higher temperatures. Each jar is identified by mnemonic code name. Jars are placed inside servo-controlled ovens maintained at prescribed temperatures throughout the test period. Thickness, volume, and mass (density = mass / volume) of each specimen are measured before swelling, and periodically after swelling under different conditions. Based on the displacement method, a special apparatus was designed and fabricated (consisting of glass beakers and graduated cylinders) for accurate volume measurements. Digital Vernier calipers are used for thickness measurements, while mass is recorded using a digital balance. Due to the toxic nature of crude oil, special care has to be taken in handling and ventilating the test area in the case of oil based elastomers. Various components of the swelling test setup are shown in **Figure 7**.

3. Study-A: Inert vs. swelling elastomer

This study is based on experiments conducted on mechanical testing and characterization of an inert (non-swelling) and a water-swelling elastomer (both belonging to the EPDM family) used for sealing purposes by a local petroleum development firm. The experiments are designed to study the effect of brine concentration, operating temperature, and sheet thickness on swelling and other properties of the elastomer. Tests were conducted to evaluate hardness, compression set (different temperatures and time periods), tensile set (different time periods), tensile properties (strength at fracture, % elongation), and swelling behavior. Swelling tests were conducted on different sample geometries, for both free and plate-mounted samples, in brine solutions of varying concentrations and at different temperatures. Total test duration was 1000 hours (about 45 days).

3.1 Experimental work

Both EPDM-type elastomers (water-swelling EPDM1, and inert EPDM2) were provided by a local petroleum development firm in the form swell packers: specified lengths of 5½ mm thickness mounted on steel pipe. Some of the elastomer sheets

Figure 7.
Different components of the swelling test apparatus.

were cautiously peeled off from the pipes. Thickness reduction for the elastomer sheets and the mounted elastomer was carried out on lathe machine and surface grinder, to match the required sample thickness. Saw-cutting and milling machines were then used to cut the mounted and free samples to requisite dimensions. Though ASTM standard sets forth a sample thickness of 6 mm, a thickness of 5 mm had to be used as the sheets had a pre-grinding thickness of 5½ mm. A dedicated die-and-punch set was used to cut the disc and ring samples. ASTM suggests rings of 18 mm OD and 1½ mm thickness; however, rings of 13 mm ID and 19 mm OD were used owing to the available punch sizes.

As discussed in published literature [5, 6] about the use of standard test methods, standards are set forth to ensure uniformity of test conditions in different locations. However, sample geometries are not always 100% binding, and a little leeway in dimensions is allowed if repeated results are consistent. Especially in the case of tensile/compressive testing, minor variations in sample dimensions do not cause any significant problems as stress is calculated per unit cross-sectional area. Forces causing the same amount of deformations are somewhat different due to the slightly different dimensions, but so is the area; net result of force per unit area (stress) remains the same. Moreover, at least three tests in each case also take care of the repeatability issue.

Reported hardness value in this work is the average of readings taken at five different locations on each sample. For the compression set test, test temperatures were room temperature (~25°C), 50°C and 80°C. For the tensile set test, ASTM standard test time is 10 min, but 10 hours and 20 hours were included to study variation of behavior more thoroughly. For tensile set and tensile properties test, ring samples were used (3 mm thickness, 3 mm width, and inside diameter of 13 mm). Disc samples (5 mm thickness, 13 mm diameter) were used for the compression set test. Hardness tests were done on 25 mm × 25 mm square samples of 5 mm thickness. Testing times were 1,000 hours (roughly 45 days) for swelling test; 10 min, 10 hours, and 20 hours for tensile set test; and 22 hours and 70 hours for compression set test.

For the swelling test, the objective was to find the volume or thickness change in salt solutions of two different concentrations (0.6%, and 20%), samples kept as three temperatures (room temperature, 50°C, and 80°C). Two types of test specimens were used. As unconfined samples (25 x 25 mm, 5 mm thickness; labeled 5 U) are not attached to any plate, and are free to swell on all sides, they give the swelling performance of free elastomer. On the other hand, samples mounted on steel plate (25 x 25 mm, 4 mm and 5 mm thickness, labeled 4 M and 5 M) are used to imitate sealing behavior of elastomer mounted on a pipe.

3.2 Results and discussion

3.2.1 Hardness

Measured hardness values (durometer Shore-A scale) ranged from 50 to 52 for the swelling elastomer EPDM1 (average 51.3) and from 59 to 62 for the inert elastomer EPDM2 (average 60.3). One reason for the slight variations in hardness values could be the peeling-off of elastomer sheets and subsequent grinding operations, giving rise to some non-uniformities. A swelling elastomer must be softer than a non-swelling elastomer (to allow water to seep in and make it swell), as confirmed by the measured hardness values.

3.2.2 Compression set

Figure 8 shows an increase in compression set CS (%) values with testing time and temperature. CS for EPDM1 (swelling type) is notably higher than that for the inert EPDM2, difference being more pronounced for longer test period and higher temperature. For the same elastomer material, 70-hour test values are higher than the 22-hour test. For EPDM1, CS values are not very different for room temperature and 50°C, but significantly higher for 80°C. We can infer from these values that permanent set would be large when the elastomer is compressed for a longer time at a higher temperature.

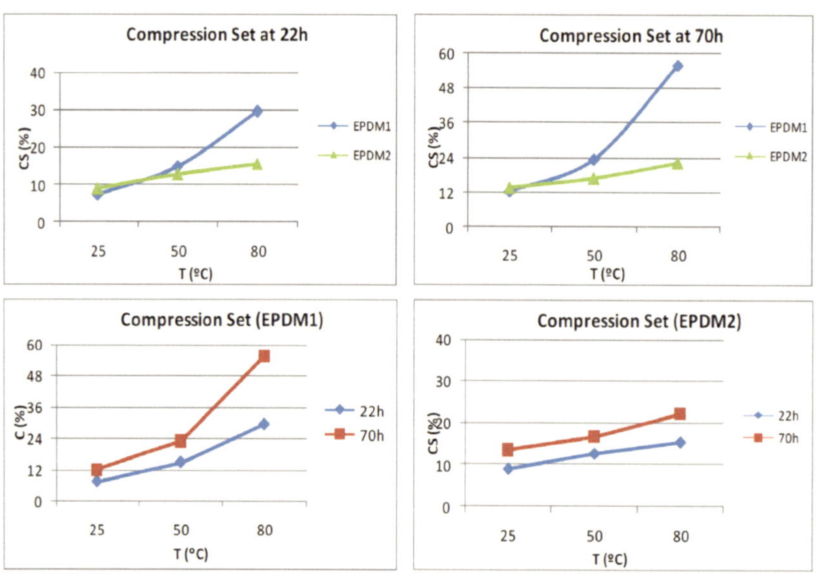

Figure 8.
Variation of compression set with testing time and temperature.

3.2.3 Tensile set

Figure 9 shows summarized values for the tensile set TS (%). As anticipated, TS values increase for longer testing time. This increase is more acute for the swelling elastomer EPDM1 than for the inert rubber EPDM2, as expected. ASTM test standard recommends a 10-min period for the tensile set test of rubbers. This appears to be too short for these much softer elastomers. Websites of many rubber vendors and manufacturers report the same testing time for both compression set and tensile set tests (22 hrs). Our in-house experiments were therefore conducted for test durations 10 min, 10 hour, and 20 hour, to have a better idea of how the behavior changes with time.

3.2.4 Tensile properties

Data from room-temperature tensile tests for both elastomer materials were converted into stress–strain diagrams; **Figure 10**. It is interesting to note that graphs for all three samples of EPDM1 nearly identical, and all curves are almost linear. This linearity makes the calculation of the slope much easier. We know that normal rubbers (like EPDM2) usually exhibit a nonlinear-elastic tensile behavior. The reason for the almost linear graphs for EPDM1 may be that swelling elastomers do not behave like normal rubbers due to their atypical cross-linking and special filler materials. For EPDM1 and EPDM2, average values of fracture stress (σ_f) were 36 MPa and 171 MPa, and of percent elongation (ε_f) were 265% and 371% respectively. Elastic modulus (E) for EPDM1 was 14.4 MPa, while it was not measured for the inert EPDM2 as the curve was nonlinear. Variation in readings for the three samples was 7%, 5% and 5% for σ_f, ε_f, and E respectively. Naturally, fracture stress and % elongation values for the far softer swelling elastomer EPDM1 are significantly lower than the inert elastomer EPDM2.

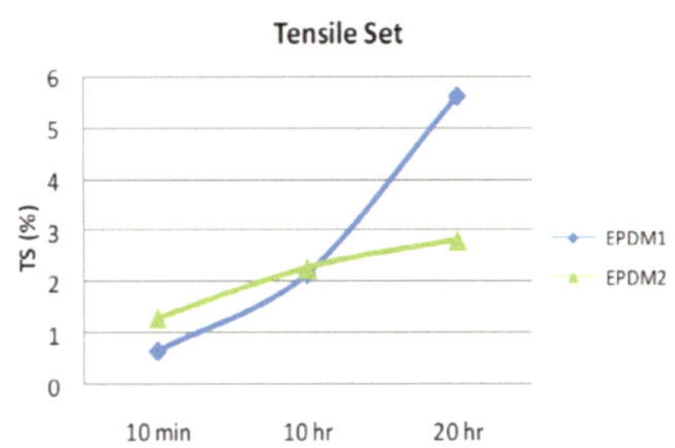

Figure 9.
Variation of room-temperature tensile set with testing time.

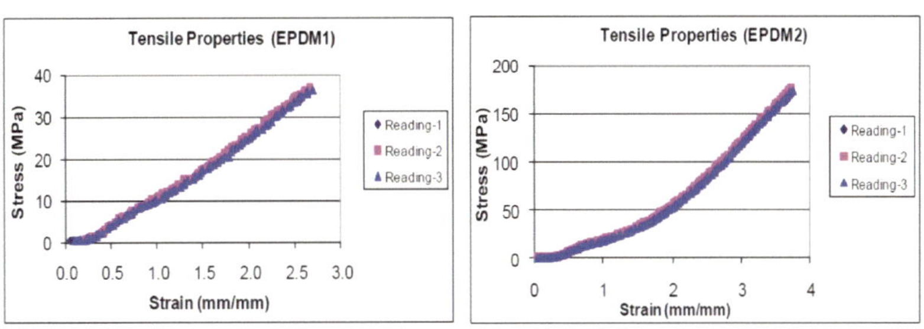

Figure 10.
Stress–strain plot from tensile test of three samples; both elastomers.

3.2.5 Swelling behavior

The most important test was of course about the swelling behavior of the elastomers. The inert elastomer (EPDM2) was of course a non-swelling type, and did not exhibit any notable changes in volume or thickness with time, as it was exposed to different saline solutions and temperatures. Graphs discussed here are therefore only for EPDM1, showing plots of swelling magnitude against time, expressed as volume change ($\Delta V\%$) and thickness change ($\Delta t\%$). Other parameters of interest are sample type and dimensions, salt concentration, and testing temperature. A systematic scheme is followed for the taxonomy (naming scheme). For example, (0.6%-80C) denotes solution of 0.6% salt concentration kept at 80°C temperature, while (4 M-50C) stands for a 4 mm-thick mounted sample tested at 50°C.

3.2.5.1 Volume change

All of the tests show that amount of swelling (in terms of % volume change) increases with increasing time. This is as expected; however, this increase is not continuous but behaves more like a step-curve: volume increases then remains constant for some time, then increases again; and so on. It is known that salt is one of the constituent materials for the swelling elastomer. As the elastomer samples are immersed in brine solutions, some salt enters into the elastomer body as water is absorbed. At the same time, small amounts of salt may also break away from the elastomer material and go into the salt solution. This two-way transport of salt means that swelling does not happen in a consistently increasing manner, but stops or even goes down for short periods of time before increasing again. Apart from the constituent materials (such as salt and other additives), one more very relevant factor in swelling elastomers is the cross-link chain density. With the breaking away of salt, some of the cross-links may be disturbed. This breaking and subsequent re-forming of cross-link chains in the elastomer may be another reason for the fluctuations in the amount of swelling [7].

If salinity value and temperature are the same, for instance 0.6% concentration and 80°C temperature (**Figure 11**), unconfined samples (5 U) exhibit higher swelling than plate-mounted samples (4 M and 5 M). It is clear that unconfined/unmounted samples are free to swell on all sides, while mounted pieces cannot swell on the surface that is restricted by the metal. If sample thickness is higher, it will obviously swell more. This is confirmed experimentally; 5 M samples show slightly more swelling than 4 M samples.

For same temperature and same sample type (eg. 4 M samples at 50°C; **Figure 11**), exposure to lower salt concentration (0.6%) generally yields higher swelling amount (volume increase). The is again an expected trend, since more concentrated and thicker solutions will not seep as much into the elastomer compare to the more dilute solutions. This will be true whether the swelling mechanism is diffusion or osmosis.

If salt concentration and sample configuration are the same (eg. 5 U samples in 20% solution; **Figure 11**), more swelling is observed at higher temperatures. This is also a natural behavior; more material expansion (swelling) and higher diffusion rates at higher temperatures.

3.2.5.2 Thickness change

Consistent with volume change behavior, it can be seen in all the graphs that sample thickness increases with swelling time. Once again, this increase is step-wise, though the fluctuation is smaller than for volume change. If salt concentration and temperature are kept constant (eg. 0.6% concentration and 50°C temperature; **Figure 11**), 5 M curve is higher than the 4 M curve, and the 5 U curve is noticeably higher than both

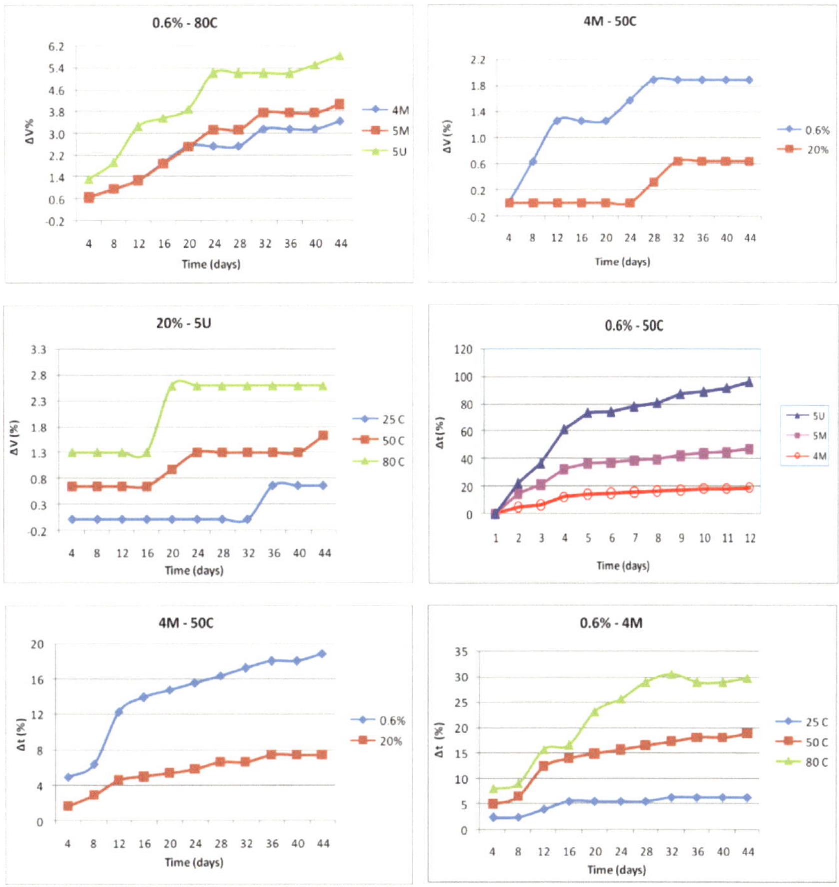

Figure 11.
Volume swelling of different sample configurations in 0.6% brine at 80°C (top left), of 4 M samples at 50°C in different salinities (top right), and of 5 U samples in 20% salinity at different temperatures (middle left). Thickness swelling of various sample geometries in 0.6% brine at 50°C (middle right), of 4 M samples at 50°C in various salinities (bottom left), and of 4 M samples in 0.6% salinity at various temperatures (bottom right).

4 M and 5 M curves. As discussed above, it is more probable that there will be larger amount of swelling (thickness change) for thicker samples, and for free (unconfined) samples as compared to the plate-mounted confined samples. When temperature and sample type are the same (eg. 4 M samples at 50°C; **Figure 11**), lower concentration (0.6%) solution yields larger thickness change. This is also natural and explained above; more swelling in dilute (less thick) solutions.

For the same salt concentration and sample configuration (eg. 4 M samples in 0.6% solution; **Figure 11**), we observe higher swelling curves for elevated temperatures, as expected. In one or two cases however, thickness seems to decrease with time. This may be some experimental error, as the general pattern is thickness increase with swelling time.

One important observation is that the total swelling amount (thickness or volume change) after the complete 45-day test period is surprisingly quite low. These samples were cut from swell packers used by the petroleum industry (elastomer sheets mounted on steel pipes). These packers were stocked in open yards for quite some time before being brought in for testing. It is well known that polymer properties are seriously affected by exposure to sun and moisture. This effect would be more prominent in the case of softer swelling elastomers. This will be discussed in more detail in Study-B.

Response of disc samples can serve as a reference, to assess the behavior of unconfined (free) elastomer. Plate samples represent the actual material response when elastomers sections are vulcanized onto steel pipes, and used as sealing elements in downhole applications. Similarly, volume swelling can be seen as a reference pattern for the swelling effect. In actual petroleum applications, thickness swelling is of actual interest as it directly relates to the sealing off of the gap between a tubular and a casing (or between a pipe and the rock formation).

3.3 Conclusions study-A

This study was primarily conducted to compare the material response of a swelling and an inert elastomer, both of EPDM type. Test plan and experimental strategy were carefully designed. Standard testing equipment was used for some of the tests, while test rigs and fixtures were designed and fabricated in-house for some other tests. All tests (except swelling behavior) were conducted in line with ASTM test standards. Inert elastomer EPDM2 was considerably harder than the swelling elastomer EPDM1 (in terms of shore-A hardness). As expected, compression set was higher for higher temperatures and longer testing times, more so for EPDM1 than for EPDM2. Room-temperature tensile set values were found to increase with testing time, again more noticeably for swelling (EPDM1) than for inert (EPDM2) elastomer. Tensile test data were converted into stress–strain graphs. Properties such as elastic modulus, tensile strength, and % elongation were extracted from the graphs. Values of fracture stress and percent elongation came out to be significantly lower for EPDM1.

Swelling tests were conducted on different sample types (unconfined and plate-mounted) for a total period of about 45 days, in brine solutions of different concentrations, and at different temperatures. In line with its inert nature, almost no swelling effect (volume or thickness change) was observed for EPDM2. For the swelling elastomer (EPDM1), volume and thickness swelling increase with increasing test temperature and decreasing salt concentration in a step-wise fluctuating fashion. Inert elastomer (EPDM2) is stronger than swelling elastomer (EPDM1), and one would assume that the sealing will last longer. However, in the case of water incursion, swelling elastomer will swell by a considerable amount, increasing the sealing pressure and providing a much better seal. All the applications of swellables in the petroleum industry (discussed in Chapter 2) are based on this novel swelling property of EDDM1 type of elastomers.

4. Study-B: Fresh vs. exposed elastomer

Zonal isolation packers and other forms of elastomer-mounted tubulars are often stacked in open yards for a long time before their deployment in wells. Elastomer properties may significantly change due to exposure (to air, sunlight, and humidity). Elastomer segments are generally covered by protective sheets; however, this wrapping can be damaged in places, exposing the elastomer to air, moisture and sunlight for long durations [8]. Some results from a comparative study of the behavior of fresh and exposed samples of an EPDM-type water-swelling elastomer are reported in this chapter.

Exposed elastomer material was provided by a local petroleum development firm, already mounted on steel pipes, ready for use as a swelling packer. Samples of fresh elastomer were supplied by a rubber development company working closely with the oilfield industry. Exposed samples are identified as EPDM1 while fresh samples are labeled as EPDM2. Chemical composition of the elastomer cannot be disclosed due to proprietary rights.

4.1 Experimental work

Elastomer properties investigated are hardness, compression set, tensile set, tensile properties, and swelling behavior. To allow for a reasonably long swelling period, the swelling test was carried out for 1,000 hours (roughly 45 days). For tensile set, test durations of 10 min, 10 hours and 20 hours were used. Compression set test was conducted for 22-hour and 70-hour periods. Test procedures and methodology has been described in detail in Section 3.1 above.

Test conditions were carefully chosen to reflect actual well environment in shallow aquifers and slightly deeper wells in regional oilfields: three temperatures (room/ambient, 50°C, and 80°C), and two salt concentrations representing low and high salinities (6,000 ppm or 0.6%; and 200,000 ppm or 20%). Two sample geometries were used for the swelling test: disc samples (25 mm diameter, 6 mm thickness), and plate samples (elastomer vulcanized on 25 mm × 25 mm steel plates). Tensile set and tensile properties tests required ring samples (3 mm thickness, 1.5 mm radial width, inside and outside diameters of 16 mm and 19 mm). Disc samples were used for compression set and hardness tests.

4.2 Results and discussion

4.2.1 Hardness

Average hardness value for the exposed elastomer (EPDM1) was 51.3 on the Shore-A scale, while that for fresh samples (EPDM2) was 57.3. This is a significant difference, indicating that hardness of a water-swelling elastomer would increase by exposure to the elements. Increased hardness (or loss of flexibility) should generally result in lower amounts of swelling. Later results corroborate this conclusion.

4.2.2 Compression set

Plots of compression set CS (%) against temperature are shown in **Figure 12** for the two test durations of 22 hr. and 70 hr. As expected, compression set increases with temperature, the increase being sharper at higher temperatures. Also, as expected, CS curve for the longer test duration is higher than that for the shorter one. This means that if the elastomer is compressed for a longer time, or at a higher temperature (or both), the amount of permanent set would be larger. As far as the comparison goes, CS values for EPDM1 are much higher than those for EPDM2. This implies that the elastomer loses elasticity due to exposure (also indicated by the hardness results), producing higher permanent set due to compression, or relative lack of springback after the release of compressive force.

4.2.3 Tensile set

Standard test duration for tensile set (TS %) test recommended by ASTM is 10 min. However, 10 hr. and 20 hr. tests were added for comparison with material data available at some of the rubber manufacturers' sites, and to study the variation pattern more thoroughly. As shown in **Figure 13**, room-temperature TS increases with increasing test period. For the 10-min test, exposed and fresh samples yield almost the same TS value. For longer testing times, curve for EPDM1 is higher than

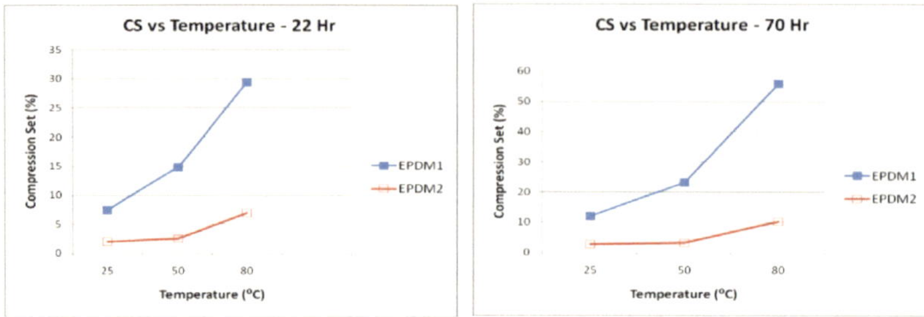

Figure 12.
Variation of compression-set with testing time and temperature.

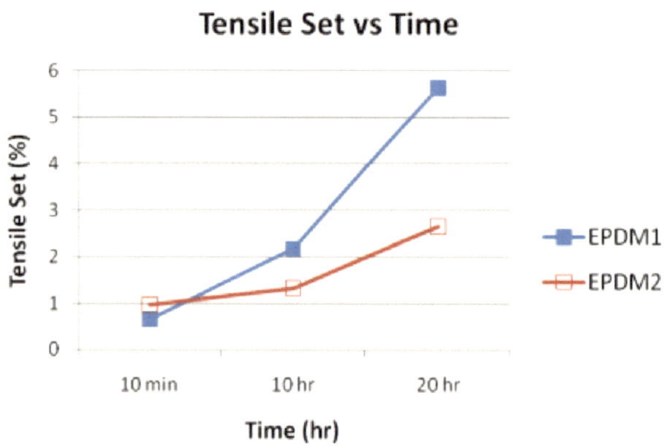

Figure 13.
Variation of room-temperature tensile set with testing time.

EPDM2, and increase in TS with time is also sharper for EPDM1. Higher permanent set under tensile loading again indicates loss of elastic recovery due to exposure.

4.2.4 Tensile properties

Figure 14 presents results of the tensile properties test for the two elastomers in the form of stress–strain graphs. As pointed out in Study-A, it is rather surprising to see that the entire stress–strain curve is almost linear for both EPDM1 and EPDM2. This would imply that the special filler materials and cross-linking used to produce swelling elastomers make them behave differently under tension as compared to normal elastomers. The fact that curves for the three samples of each elastomer are almost identical, together with the near-linearity of the curves makes it very convenient to calculate tensile properties, especially the elastic modulus (slope).

As summarized in **Table 1**, average fracture stress and elastic modulus for exposed samples are significantly higher than that for fresh ones. Percent elongation shows an opposite trend. This reinforces the previous results; exposure reduces the softness of the elastomer, resulting in lower flexibility (percent elongation) and higher fracture stress.

4.2.5 Swelling behavior

Being a water-swelling elastomer, the most crucial test was the determination of swelling response when the elastomer is immersed in saline water at different temperatures. In **Figures 15–18**, amount of swelling (volume change ΔV % and

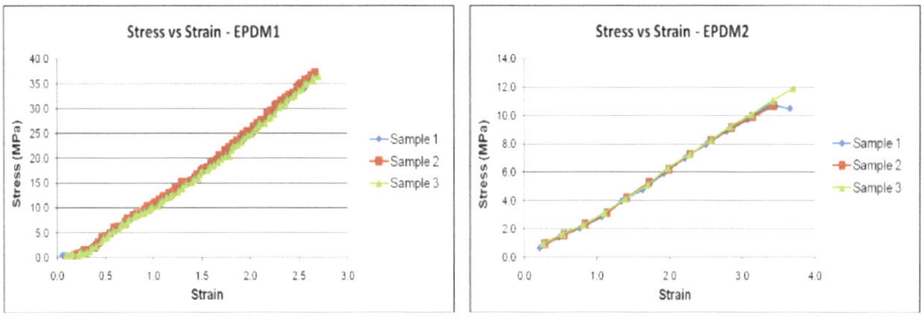

Figure 14.
Stress–strain plots from tensile test of three samples of fresh and exposed elastomer before swelling.

Elastomer type	Fracture stress (MPa)	Percent elongation (%)	Elastic modulus (MPa)
EPDM1 (Exposed)	35.96	264.47	14.40
EPDM2 (Fresh)	11.09	359.23	3.22

Table 1.
Average tensile properties of exposed and fresh elastomer.

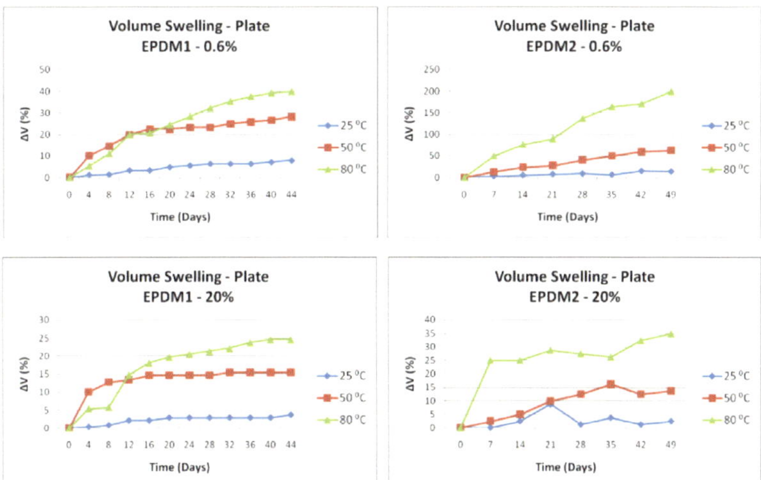

Figure 15.
Volume swelling of plate samples; low and high salinities; different temperatures.

thickness change Δt %) is plotted against time for different sample types and test conditions.

4.2.5.1 Plate samples

Figures 15 and **16** show the variation of swelling with time for plate samples. It is clear from all of the graphs that more swelling occurs when samples are kept under water for longer duration, as expected. As observed in Study-A, this increase in volume (or thickness) with time does not progress smoothly, but happens in a fluctuating manner. Swelling increases, then remains constant for some time, then increases again, and even decreases a bit at times. Possible reasons (buildup/reduction of salt content, and changes in cross-link structure) are the same.

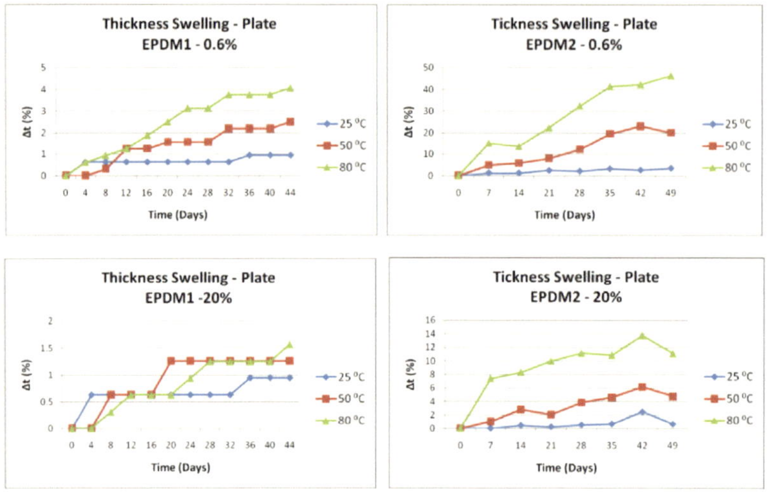

Figure 16.
Thickness swelling of plate samples; low and high salinities; different temperatures.

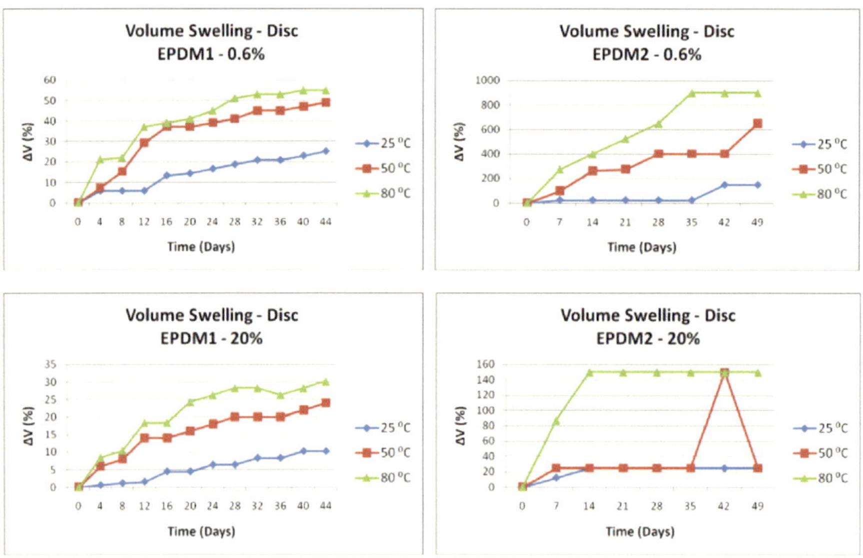

Figure 17.
Volume swelling of disc samples; low and high salinities; different temperatures.

Maximum swelling (volume or thickness) occurs for both elastomer types at a temperature of 80°C and under 0.6% salt concentration. Obviously, like most of the other materials, elastomers expand more with increasing temperature. Expanded pores allow more water to soak into the material, resulting in higher swelling. It is also natural that an elastomer would swell more in diluted solutions than in higher-concentration solutions, whether swelling happens due to diffusion or due to osmosis.

Volume swelling percentage is evidently much higher than thickness swelling; thickness change represents swelling in only one direction, while volume change corresponds to swelling from all exposed surfaces. Under the same conditions of salt concentration and temperature, fresh samples show substantially higher swelling than exposed ones (200% volume change compared to only 40% for 0.6%-80°C

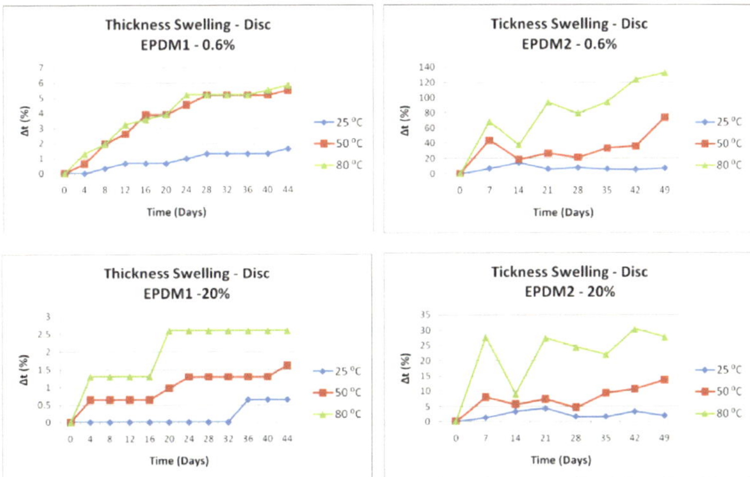

Figure 18.
Thickness swelling of disc samples; low and high salinities; different temperatures.

condition, for instance). This is in line with all earlier observations; extended exposure to sun, wind and moisture reduces elastomer flexibility and increases its hardness; harder and less elastic material naturally exhibits reduced amounts of swelling.

4.2.5.2 Disc samples

Volume and thickness swelling of disc samples is plotted against swelling-time in **Figures 17** and **18**. Like plate samples, discs also demonstrate the fluctuating swelling pattern. Also, in a similar manner, maximum swelling is observed at the highest temperature and the lowest salt concentration. As before, amount of volume swelling is far higher than thickness swelling. Most importantly, once again, fresh samples undergo noticeably higher swelling than exposed ones (900% volume change compared to only 55% under the 0.6%-80°C condition). Reasons for these observations for disc samples are the same as those described above for plate samples.

If we compare the swelling response of disc samples against plate samples, we notice a huge difference (for example, 900% volume change as against 200%, or 130% thickness change in comparison with 48%, under the same salinity and temperature conditions of 0.6% and 80°C). As explained earlier, disc samples are free to swell from all sides, while swelling of plate samples is restricted from one major surface; thus the sizeable difference.

4.3 Conclusions study-B

Comparison between material response of two sample sets of an EPDM-type water-swelling elastomer has been carried out through material characterization experiments, in particular the study of swelling behavior. One set of samples was cut from fresh elastomer, and the other from elastomer exposed to air, moisture, and sunlight. Shore-A hardness of exposed elastomer samples (EPDM1) was notably higher than that of fresh samples (EPDM2), indicating loss of flexibility with exposure. Compression set was found to increase with increasing temperature and testing time, CS values of exposed samples being significantly higher than fresh ones. Room-temperature tensile set values of the two sample types were almost the same for short-duration test (10 min), but were considerably higher for exposed elastomer after longer-duration tests. Both compression set and tensile set values suggest that permanent set (or lack of springback) increases with exposure to the elements. Tensile properties test data surprisingly yielded almost linear stress–strain graphs for both sample sets, as against highly nonlinear graphs for most rubber-like materials. Average values of fracture stress and elastic modulus for the exposed elastomer were clearly higher, while percent elongation was lower, again implying that exposure reduces softness and flexibility of the elastomer. Rather than increasing steadily with time, swelling response showed a fluctuating trend for both elastomers. Higher amount of swelling is generally observed for higher temperatures and lower salt concentrations. Under the same conditions of temperature and salinity, fresh elastomer samples exhibited far more swelling (percent volume or thickness change) than the exposed samples. This strengthens the observation that elastomers lose suppleness if exposed to sun and moisture, etc. for extended periods of time.

5. Study-C: Swelling under water, oil, and acid

Well stimulation is the name given to techniques that are performed to increase or restore production. If a well initially exhibits low permeability, stimulation is used to start production from the reservoir. In other cases, stimulation is used to improve permeability and flow from an already existing well that has become under-productive. Acid induction, or acidizing, is used to either stimulate a well to improve flow or to remove damage. In matrix acidizing, acid is injected into the well, penetrating the rock pores at pressures below fracture pressure. The acid dissolves the sediments and mud solids within the pores that are inhibiting the permeability of the rock, thereby enlarging the natural pores of the reservoir and stimulating the flow of hydrocarbons, but this acid does not react with the hydrocarbons [9, 10].

Swelling beahavior of inert and swelling elastomers, and fresh and exposed elastomers was discussed in study-A and study-B above. Results from an experimental investigation are presented in this section about the behavior of two commercial elastomers, one water-swelling and one oil-swelling, with and without one-day acid exposure. Out of the one-month total testing time, one set of samples (for each elastomer type) was tested under acid for one day, and the other set without any acid exposure. Changes in volume swelling, thickness swelling, and hardness of elastomer samples were recorded at various prescribed times during a one-month swelling period. Selection of test parameters such as water salinity, temperature, acid concentration, and type of crude oil was based on actual field conditions in target regional oil wells.

5.1 Experimental work

Two sample geometries (disc and plate) were used for the two elastomer materials (one water-swelling, and one oil-swelling). All samples were provided by a regional oilfield operator, name and number of the elastomer not to be disclosed due to confidentiality reasons. Salt-water solutions of 35000 ppm (3.5%) and 85000 ppm (8.5%) strength were used to test the water-swelling elastomer, while crude oil from the field was used to swell the oil-based elastomer. The oil is classified as 'light crude oil, which is a liquid hydrocarbon that may contain up to 0.6 vol% benzene and other light aromatics and 0.8–1.5 wt% sulfur compounds, and has a viscosity of 200–450 cp. Same test temperature of 60°C was employed for both elastomers. To replicate conditions of acid induction in the actual wells, samples were removed from water or oil after 3 days of swelling, and placed in 15% HCl solution for one day. They were then returned back to swell in water or oil. Measurements were taken on day-zero (no swelling) and after 1, 3, 4, 7, 15, and 31 days of swelling.

All reported results are average of readings from 3 samples; **Figure 19**. After measurements, plates and discs were returned back to their jars. In earlier studies, hardness samples were discarded after each measurement and were not placed back into brine or oil. There was a concern that small puncture marks made on the surface by the durometer indenter may affect future swelling to some degree. However, this necessitated a very large number of samples. For the current study, the rubber company opted for a limited number of samples, each one to be re-used after hardness measurements. This may result in minor deviations from expected swelling behavior.

5.2 Results and discussion

Results are presented in graphical format and behavior patterns are analyzed for swelling under water, under oil, and under water or oil with intermediate acid exposure for one day. Variations discussed are changes in volume and thickness swelling, and in hardness values. To have a general idea about the variation in data, all swelling related graphs include error bars based on a 95% confidence interval. Though results for all cases were carefully tabulated in detail and behavior patterns plotted, only a few are presented here for illustration. **Figure 20** shows physical exhibits of the amount of swelling (volume change, thickness change) and surface texture of swollen elastomer in some cases.

5.2.1 Volume change

Percentage change in volume due to swelling of disc samples against swelling time (number of days) in 35000 ppm and 85000 ppm brine solutions is shown **Figure 21**, with and without acid exposure. As expected, and as found in earlier studies [7, 8, 11], amount of swelling is higher in lower-strength brine than in higher higher-concentration one. Effect of acid exposure on swelling performance is not so straightforward. In 8.5% brine, swelling slows down during day-4 (when samples were removed from water and put into HCl solution), but does not do so in 3.5% salt solution. After this one-day acid induction, swelling amount steadily increases in both cases when the elastomer samples are placed back into the brine. Total acid-affected volume swelling at the end of the one-month period is almost the same

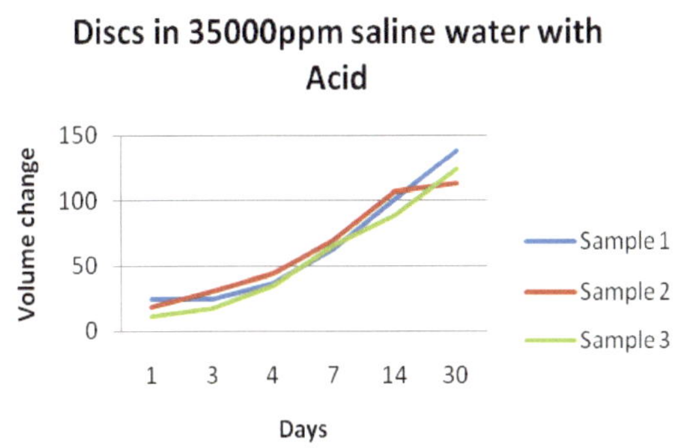

Figure 19.
A 3-sample graph of volume swelling against time for disc samples in 35000 ppm saline water.

Figure 20.
Progressive swelling and surface texture of disc and plate samples in 35000 ppm saline water (above) and oil (below).

(with and without acid introduction) in 3.5% solution, but significantly higher in 8.5% solution. Plate samples of water-based elastomer exhibit similar behavior.

Volume change with swelling time for plate samples of the oil-based elastomer is shown in **Figure 21** (bottom). As for the water-swellable elastomer, samples keep up the swelling trend after the one-day acid exposure. Swelling goes down during acid-induction (day-4), but shows a somewhat erratic trend (increasing, decreasing, and increasing again) for the no-acid-exposure case. Total volume swelling after acid induction is significantly higher. Disc specimens also show similar swelling trend, though less irregular than plate samples. Higher amount of post-acid swelling for both water and oil-based elastomers hints at extra softening of the elastomer during acid exposure.

5.2.2 Thickness change

Variation of thickness for plate samples against swelling time (number of days) in 35000 ppm and 85000 ppm salt solutions is shown in **Figure 22**, with and without acid induction. As observed earlier, higher salt concentration leads to lower amount of thickness swelling. Total thickness change (one month swelling time) with acid exposure is a little higher in both cases. Change in thickness swelling against number of days for plate samples of the oil-swelling elastomer is shown in **Figure 22** (bottom). End-of-month thickness swelling is significantly higher in the case of acid exposure. Also, thickness swelling increases more rapidly after the one-day acid induction. Disc samples behave in a similar manner. Significantly higher amount of post-acid volume and thickness change indicates that there is more softening due to acid exposure in oil-swelling elastomer than in the water-swelling material. This observation is confirmed by the hardness results presented later.

5.2.2.1 Mechanism of swelling

Oil swelling elastomers are predominantly based on EPDM (ethylene propylene diene monomer, M-class) type of rubbers. Rubbers that have a saturated chain of the polymethylene type are categorized as M-class according to ASTM standard

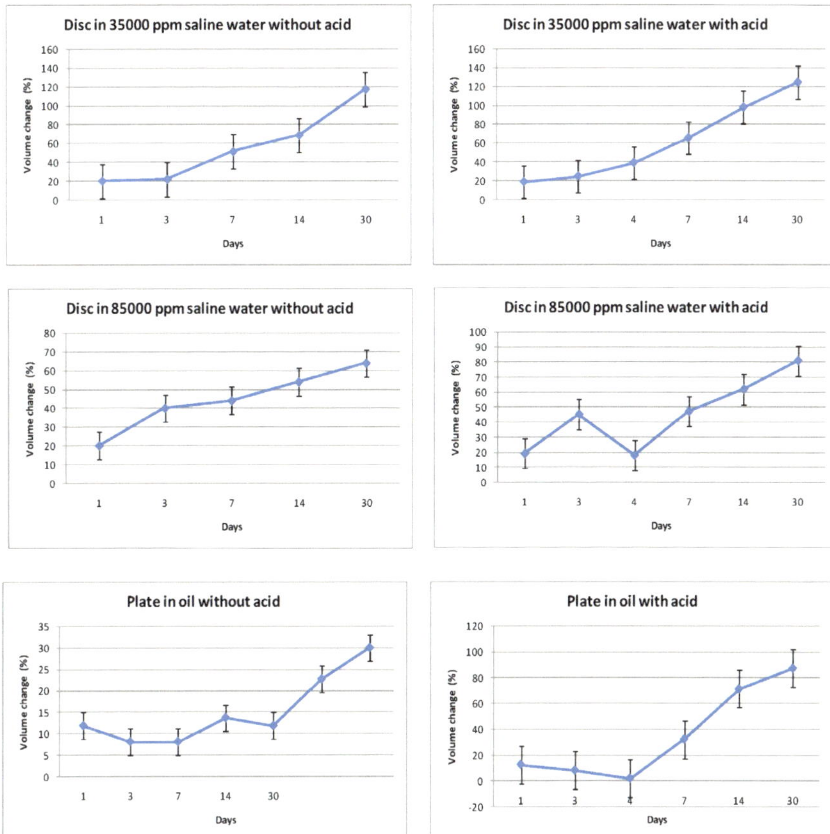

Figure 21.
Volume change (%) against swelling time, with and without acid exposure; disc samples in 3.5% and 8.5% brine solutions (top and middle); plate samples in oil (bottom).

D-1418. Materials with similar values of solubility parameter are likely to be miscible. Both EPDM and petroleum are nonpolar molecules, and are miscible together. It should be pointed out however that oil swelling in elastomers is not a process of dissolving the rubber. For instance, solubility parameter of diesel fuel is the 7.5–8.5 range, while that of EPDM is around 8.0 [12]. That is why uncured EPDM can be dissolved, while cured EPDM swells in crude oils. On the other hand, solubility parameter of NBR (nitrile-butadiene rubber) is about 9–10.5. Because of this significant dissimilarity, NBR swells very little in a hydrocarbon. Another important factor is the polymer-solvent interaction parameter [13] used by Flory and Huggins to represent the change of Gibbs free energy. Value of this interaction parameter for cross-linked EPDM and hydrocarbon is slightly greater than 0.5. EPDM type elastomers therefore swell rather than dissolve in crude oils. Interaction parameter value for oil swelling depends on various factors such as type of base elastomer, cross-linking density, type of oil, viscosity of oil, and operating temperature [14].

Water swelling elastomers are mainly compounded from nitrile or hydrogenated nitrile rubbers [15]. Super absorbent polymers (SAP), organic/inorganic salts, and/or any saline materials are a vital ingredient to boost the absorption of water into the rubber matrix to cause swelling. As an example, sodium polyacrylate is a polyelectrolyte SAP (widely used in disposable diapers) and swells due to osmotic pressure effects [16]. The anionic charge on the pendant groups of the polymer chain needs to be balanced by cationic counter ions ($Na+$). To equalize the chemical potentials of the counter ions (or to maintain thermodynamic equilibrium), water

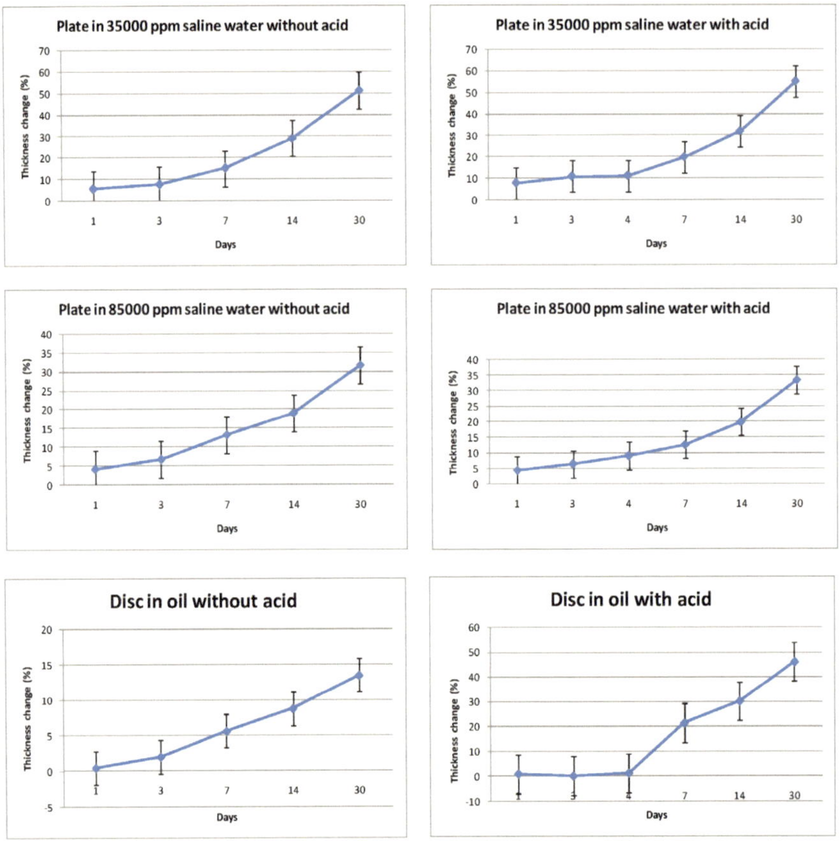

Figure 22.
Thickness change (%) against swelling time, with and without acid exposure; plate samples in 3.5% and 8.5% brine solution (top and middle); disc samples in oil (bottom).

migrates into the elastomeric matrix, leading to its swelling. The amount of swelling of the elastomeric matrix depends on the balance between the stretchability of the polymer network and the osmotic pressure effects of the counter ions. Elastomer stretchability is a function of cross-linking density, physical properties of the elastomer, and environmental temperature. Osmotic pressure depends on temperature, and the diffusion rate therefore increases with temperature, due to the higher movement of molecules in general [14].

5.2.3 Disc vs. plate samples

Amount of volume or thickness swelling for disc samples is significantly higher than that for plate samples in all cases (3.5% brine, 8.5% brine, oil), with and without acid exposure. This is both intuitively expected and borne out by earlier studies. Due to gluing/vulcanizing of the elastomer onto a metal base, plate samples are not free to swell on one major surface; disc samples have no such restrictions. Percentage change in swelling is therefore much lower for plate samples in comparison with disc samples.

5.2.4 Water-swelling vs. oil-swelling elastomer

Percent change in volume and thickness for water-swelling elastomer is generally far higher than that for oil-swelling elastomer. However, in one case, oil-swelling elastomer shows an outlier behavior: higher volume swelling for disc in oil as compared to disc in water (for acid-affected samples). Field engineers can use this general trend to form an important operational policy: *to generate the desired sealing pressure, annular separation between the packer and the casing (or formation) for oil-swelling elastomers needs to be much smaller than that for water-swelling elastomers.*

5.2.4.1 Outlier behavior

The general pattern observed in all graphs is that swelling increases with longer exposure to the swelling medium, but the increase is not always uniform. This is in line with earlier works by the authors and by other researchers. However, in one or two instances (such as in **Figure 21**), swelling stops at a certain level, or goes down a bit, before increasing again. Water transport is a two-way mechanism; from the brine into the elastomer, and back from the swollen elastomer into the salt solution. In general, due to the prevalent concentration difference, there is a net movement of water into the elastomer, causing increase of swelling with longer exposure to water. Under certain conditions, and for brief intervals of time, the process can reverse (as in **Figure 21**), causing de-swelling of the elastomer. Very rarely, the contacting fluid might also leach out soluble constituents of the elastomer's recipe, to reduce test sample dimensions [17].

5.2.5 Hardness change

Reduction in hardness of an elastomer due to progressive swelling after prolonged exposure can significantly affect seal integrity. Variation in hardness against swelling time for disc samples in salt solutions of 3.5% and 8.5% concentration is plotted in **Figure 23**, with and without exposure to acid. Durometer hardness of original elastomer samples was around 53–55 on the Shore-A scale. Within 3 days of swelling, hardness sharply drops down to around 30, and then decreases more gradually with further swelling. Final hardness after one-month exposure is almost the same in both 3.5% and 8.5% brine, with and without acid induction. **Figure 23** (bottom) summarizes the hardness behavior against number of days of swelling for plate samples of oil-based elastomer. Hardness of the unswelled samples was about 58 Shore-A. In this case, it takes much longer for a significant decrease in hardness; one to two weeks of swelling. Also, acid injection reduces the hardness value significantly. Lower hardness values and larger amounts of volume and thickness change are consistent with each other. Acid exposure results in more softening of oil-based elastomer, which leads to higher amounts of swelling.

5.2.5.1 Drop in hardness and seal integrity

It can be easily observed that hardness behavior of both water-swelling and oil-swelling elastomers is quite similar, showing almost similar and significant hardness reduction the first few days (with and without acid). Large amount of swelling (as reported above for volume and thickness change) is good as it generates large sealing pressure. However, a practical question about long-term integrity is very important. How will the elastomer seal (with relatively soft swollen elastomer elements) perform over extended periods of time if subjected to high pressure differential? Reports from oilfields using swelling elastomer applications are encouraging in general. Enough sealing pressure is generated by the large amount of swelling to offset weakening of the material due to softening. Most of the applications work satisfactorily for at least a few years. However, it is rather impossible to arrive at

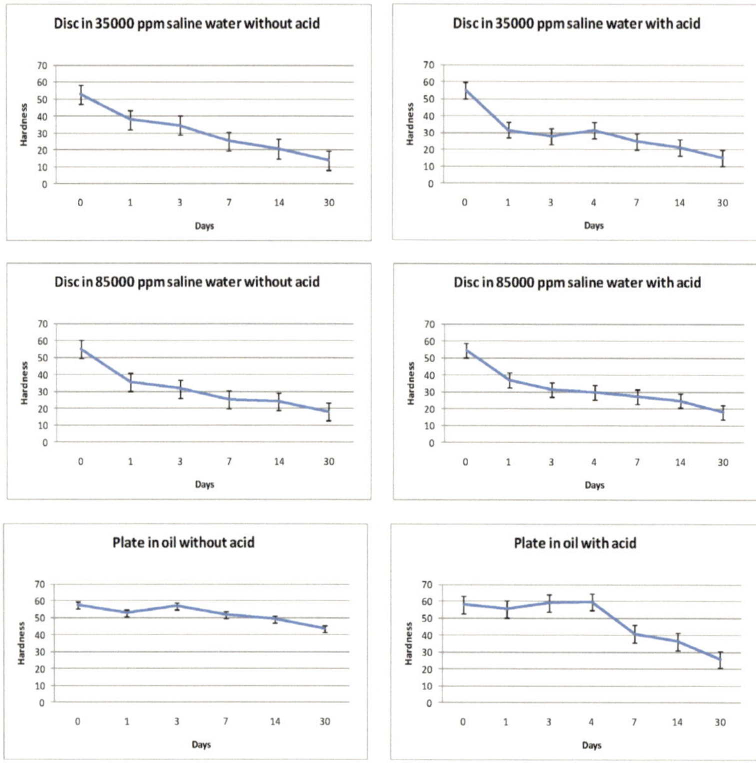

Figure 23.
Effect of swelling on hardness, with and without acid exposure; disc samples in 3.5% and 8.5% salt solutions (top and middle); plate samples in oil (bottom).

any reliable conclusion without actual long-duration tests under different conditions (temperature and pressure) and in different swelling media (saline water, oil, acid). An experimental facility for longevity testing of different water and oil-swelling elastomers has been designed and constructed by the authors at Sultan Qaboos University, Muscat, Oman [18]. Testing is in its fifth and final year by now. Conclusions from this one-of-a-kind ongoing study will be quite illuminating.

5.3 Conclusions study-C

A series of experiments was designed and carried out to investigate the effect of swelling on material response of a water-swelling and an oil-swelling elastomer, with and without acid exposure. Three swelling media were used: salt solutions of 35000 ppm and 85000 ppm concentration, crude oil, and 15% HCl. Testing temperature was 60°C in all cases. One set of elastomer samples (both water and oil-swelling) was placed into acid solution for one day after 3 days in the original medium (brine or oil), and was then put back into the same medium for the remainder of the one-month period. Measurements for volume, thickness, and hardness were done before swelling, and after 1, 3, 4, 7, 15, and 31 days of swelling. All readings were recorded for three samples and averaged out. As expected, lower concentration brine leads to higher amount of swelling. Being free to swell in all directions, changes in volume and thickness due to swelling are higher for disc samples as compared to plate samples, in water and in oil. For the one-day when samples are exposed to acid, amount of swelling generally goes down a little, before picking up again on exposure to the original media. Both elastomer types exhibit a sharp decrease in hardness during the first few days of swelling, and then hardness gradually becomes stable for the rest of the swelling period. Acid exposure leads to higher amount of swelling and lower hardness values for both water and oil based elastomers during the remaining post-acid swelling time. Results from this investigation can be used by engineers and practitioners in oil and gas fields for pre-qualification and appropriate selection of swelling elastomers to suit targeted field conditions. These results can also be used as input parameters for modeling and simulation of swelling elastomer seal performance, leading to improvements in design and manufacturing of swell packers.

Author details

Sayyad Zahid Qamar[1]*, Maaz Akhtar[2] and Tasneem Pervez[1]

1 Mechanical and Industrial Engineering Department, Sultan Qaboos University, Muscat, Oman

2 Mechanical Engineering Department, N.E.D. University of Engineering and Technology, Karachi, Pakistan

*Address all correspondence to: sayyad@squ.edu.om

References

[1] Akhtar M, Qamar SZ, Pervez T, Al-Jahwari FK (2018) "Performance Analysis of Swelling Elastomer Seals," Petroleum Science and Engineering, 165 (2018), p 127-135

[2] ASTM D2240-05 (2007) "Standard Test Method for Rubber Property – Durometer Hardness," ASTM International, West Conshohocken, PA

[3] ASTM D395-03 (2007) "Standard Test Methods for Rubber Property — Compression Set," ASTM International, West Conshohocken, PA

[4] ASTM D412-06a (2007) "Standard Test Methods for Vulcanized Rubber and Thermoplastic Elastomers — Tension," ASTM International, West Conshohocken, PA

[5] Davis JR (2204) Tensile Testing, 2nd edition, ASM International, Materials Park, Ohio, USA

[6] Kuhn H, Medlin D (2000) ASM Handbook Volume 8: Mechanical Testing and Evaluation, ASM International, Materials Park, Ohio, USA

[7] Qamar SZ, Pervez T, Akhtar M (2016) "Performance Evaluation of Water-Swelling and Oil-Swelling Elastomers," Journal of Elastomers and Plastics, 48 (6), 2016, p 535-545

[8] Pervez T, Qamar SZ, van de Velden M (2012) "Comparison between Fresh and Exposed Swelling Elastomer," *Journal of Elastomers and Plastics*, 44 (3), May 2012, p 237-250

[9] Medina E, Sierra J, Garcia A, Gleaves J, Mendez J (2015) "Optimization of Matrix Acidizing With Fluids Diversion in Real-Time Using Distributed Temperature Sensing and Coiled Tubing," SPE Paper # 173686, SPE/ICoTA Coiled Tubing and Well Intervention Conference and Exhibition, 24-25 March 2015, the Woodlands, Texas, USA

[10] Stringfellow WT, Camarillo MK, Domen JK, Shonkoff SBC (2017) "Comparison of chemical-use between hydraulic fracturing, acidizing, and routine oil and gas development," PLoS ONE 12(4): e0175344

[11] Qamar SZ, Pervez T, Akhtar M, Al-Kharusi MSM (2012) "Design and Manufacture of Swell Packers: Influence of Material Behavior," Materials and Manufacturing Processes, 27 (7), 2012, p 727-732

[12] Mutelet F, Ekulu G, Solimando R, Rogalski M (2004) "Solubility Parameters of Crude Oils and Asphaltenes," Energy Fuels, 18 (3), 2004, p 667-673

[13] Orwoll RA, Arnold PA (2007) "Polymer–Solvent Interaction Parameter χ," p 233-257, chapter in book Physical Properties of Polymers Handbook, edited by JE Mark, Springer, 2007

[14] Kim BJ (2014) "Studies of swellable elastomers for oilfield services," *Rubber World*, Dec (2014)

[15] Polgar LM, Fallani F, Cuijpers J, Raffa P, Broekhuis AA, van Duin M, Picchioni F (2017) "Water-Swellable Elastomers: Synthesis, Properties and Applications," Reviews in Chemical Engineering, 35 (1), p 45-72

[16] Kleverlaan M, van Noort RH, Jones I (2005) "Deployment of Swelling Elastomer Packers in Shell E&P," Paper # SPE-92346, *SPE/IADC Drilling Conference*, 23-25 February, Amsterdam, Netherlands

[17] Campion RP, Thomson B, Harris JA (2005) "Elastomers for fluid

containment in offshore oil and gas production: Guidelines and review," Research Report 30, HSE Books, 2005

[18] Qamar SZ, Pervez T, van de Velden M, Sanchez FJ (2012) "Design and Fabrication of Test Facility for Longevity Testing of Elastomer Seals," *ASME-2012 International Mechanical Engineering Congress and Exhibition* (IMECE-2012), Houston, Texas, 9-15 November 2012

Chapter 4

Experimental Setup for Swellable Elastomers in Cased and Open Holes

Sayyad Zahid Qamar, Maaz Akhtar and Tasneem Pervez

It is an essential characteristic of experimentation that it is carried out with limited resources, and an essential part of the subject of experimental design is to ascertain how these should be best applied; or, in particular, to which causes of disturbance care should be given, and which ought to be deliberately ignored.

Sir Ronald A Fisher

Abstract

A full scale experimental setup was designed and commissioned for testing of swelling elastomer seals against a casing (cased hole) and formation (open hole). Actual replicate of wellbore was designed with varying inside diameters and roughness to reproduce the effect of actual formation. The Dynaset packer mounted on a 7-inch tubular was allowed to swell against a 9–5/8-inch casing, while the fast swell packer mounted on a 9–5/8-inch tubular was allowed to swell against the 12–1/4-inch replicated well bore. This one-of-its-kind test setup can demonstrate the way the elastomers swell out and fill the asperities against smooth outer casing (cased hole) or against rough wellbore surface (open hole). Dismantling of the test setup midway through the testing scheme revealed a severely dimpled surface of the swelled elastomer.

Keywords: demonstration setup, elastomer swelling, cased hole, open hole

1. Introduction

In the petroleum industry, packers refer to the components/products which are used to isolate one section in a formation from others, or to isolate the outer section of a production tubing from the inner section, which may be a casing or liner or well-bore itself [1]. They are also called *mechanical packers* because they remain in an oil well during its production life, and are set by some form of tubing movement. In a typical well, a packer is installed close to the bottom end of the production line, just above the top perforations [2]. If a well is connected to multiple reservoir zones, then it is also used to isolate the perforations for each zone. It may also be used to isolate sections of corroded casing, casing perforations, casing protection from undesirable fluids, and complete shut-off of high water cut-off zones. These production packers are further classified into *permanent* or *retrievable* packers [3]. Permanent packers are low in cost and provide better seal, while retrievable packers are expensive but re-usable after repair, and have lower sealing capability [4, 5].

Temperature and pressure can cause changes in the expansion rate of production tubing and packer leading to the development of tensile or compressive or no force conditions [6].

Till the innovative use of swellable elastomers, the most commonly used packers were the ones mentioned above. A swellable packer is much simpler than a mechanical packer because there are no moving parts and it requires no special services. Like a casing, the swellable packers are lowered to the desired depth and allowed to swell before production or injection operations begin [7]. These elastomers swell when they come in contact with different types of fluids and are categorized based on the fluid, such as oil-based swellable elastomer, saline-water based swellable elastomer, etc. The swelling amount of such elastomer depends on the fluid chemistry and the temperature at swell. Typically, the elastomers are vulcanized on a pipe to make a swellable packer. When it comes in contact with a fluid, it starts to swell. The swelling continues until the internal stresses inside the elastomer reach equilibrium, and creates a differentially sealing annular barrier, hence completing the sealing [8].

In field applications, different types of swellable elastomers are used at different depths of the same well, or in different wells. Extensive amount of research has been carried out for material and mechanical characterization of swellable elastomers in its virgin or vulcanized form as well as in bore-hole conditions. The characterizations had been done for some of the elastomers under different conditions of temperature and salinity, using laboratory-size elastomer samples [9–12]. However, the uncertainty remains about the effectiveness of actual sealing done by the swellable packers. Is the packer really effective in terms of providing the needed pressure to keep the unwanted fluid away from seeping into the production tubing? Do these swellables completely cover the random profile of formation in open bore-holes or the roughness of a casing in cased holes? However, no tests have been carried out yet in a replication of the real oilfield scenario: elastomer seals mounted on tubulars in cased or open holes, with or without cementing. This chapter explains the performance of a mix of swellables under these scenarios through the development of an experimental demonstration unit.

This demonstration unit is designed and fabricated using inert and swelling elastomer seals mounted on actual petroleum pipes/tubulars, and housed in a concrete block replicating the open-hole and a casing replicating the cased-hole. The inner-most pipe is a swell packer comprising inert and swelling elastomer segments sealing against a casing pipe, which in turn uses another set of swelling elastomers to seal against the formation, in the form of a concrete block with layers of varying roughness as would be the case in a real formation [13].

The main objectives of this demonstration-cum-test unit were:

a. Life-size demonstration of how a swelling elastomer packer provides sealing against the formation in a well hole.

b. Life-size demonstration of how a swelling elastomer packer provides sealing against a casing in a well hole.

c. Testing the actual behavior of fast-swell elastomer (E1) in near-ambient conditions in a shallow aquifer.

d. Testing the actual behavior of a swell packer elastomer (E2) in near ambient conditions in a shallow aquifer.

e. Testing and demonstration of whether swelling elastomers completely seal off irregular boreholes and geometrical aberrations.

2. Design of experimental unit

Many possible designs were considered for this experimental unit. The following set of criteria reduced the possible designs to only a few:

a. The experimental unit must replicate the formation roughness of typical on-shore oil-field as closely as possible.

b. The casing, production tubing, and swell packers (along with the spacing between these) had to be the same as used in an actual oil-well.

c. The spacing between open-hole and packer, and cased-hole and packer must be filled with fluids that are compatible to specifications of the swellable elastomers used.

d. Swelling behavior of the elastomer must be clearly visible and recordable during the swelling process.

e. The experimental unit must be able to de-swell the elastomer after complete swelling and re-swell again to see the effect of the swell/de-swell/re-swell cycle.

Based on the above criteria, the final design consisted of a half (cutaway) concrete block replicating the formation. Symmetrically cutaway sections of two concentric swell packers were housed inside this concrete half-block. The sealing elements between the two packers and between the outer casing and formation (concrete) were actual inert and swelling elastomers. The semi-circular cavity in the concrete block was to have slight variations in diameter and roughness to reproduce realistic formation conditions. The setup was to be enclosed from the top and

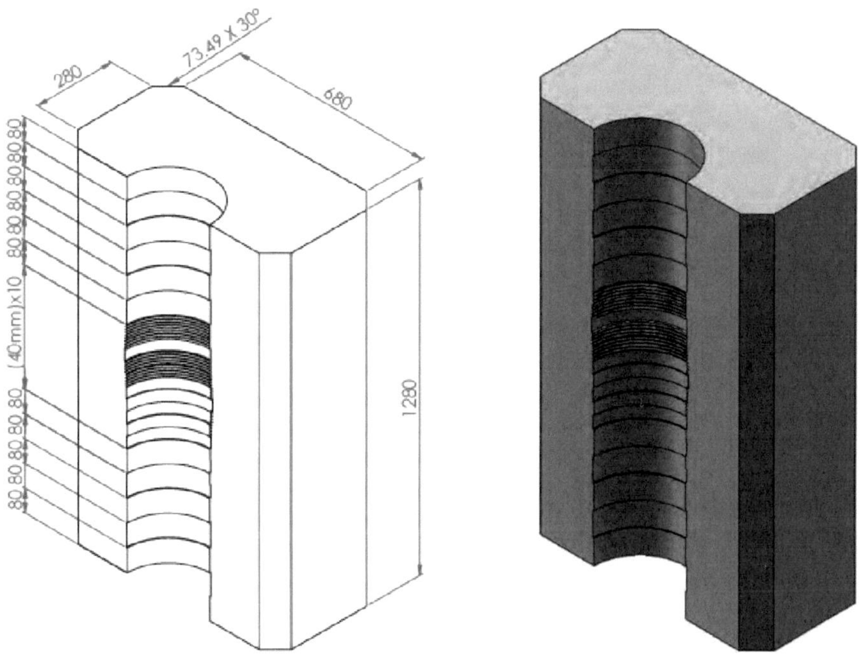

Figure 1.
Concrete block with layers of different roughness to replicate formation geometry.

front with high-strength plexi-glass, to be able to withstand high pressures, and to provide a clear view of the internal arrangement and swelling behavior. The final configuration of the experimental unit is shown in **Figures 1** and **2**.

Major specifications of the final design include the following: Concrete half-block: height 1280 mm; width 827 mm; thickness 427 mm. Semicircular hole in concrete half-block consisting of 80 mm deep sections of varying roughness. Diameter of semicircular hole varying from 311 to 320 mm as per an actual oil-well. Inside packer consisted of two elastomer segments of inert and swellable type; elastomer on 177.8 mm tubular swells against 244.475 mm casing. Outside packer consisted of fast-swell elastomer mounted on 244.475 mm casing that swells against 311.5 mm formation (concrete block). Test conditions included fresh water with almost no salinity, ambient temperature of about 20-degree Celsius, and atmospheric pressure (representing shallow aquifers in actual oil wells).

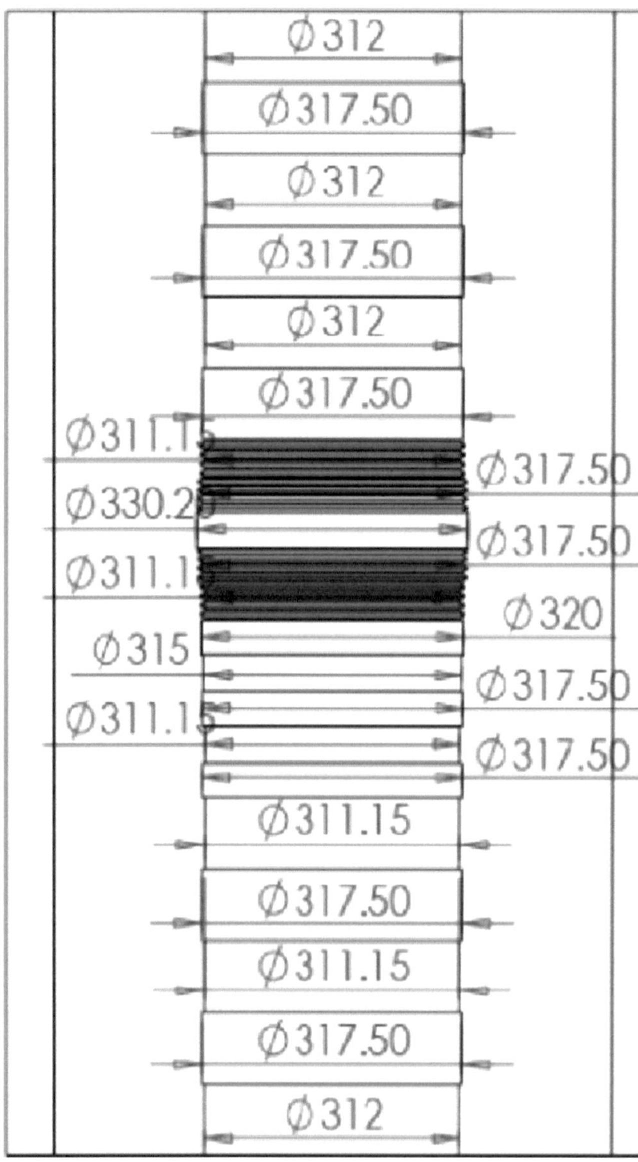

Figure 2.
Different diameters and roughnesses to replicate variations in the drilled formation.

3. Fabrication/assembly of experimental unit

In consultation with oilfield engineers, and following the recommendations of concrete requirement for oil-well cementing jobs, a special concrete mixture was prepared to provide sufficient strength, durability and relatively quick curing time. In order to develop different inside diameters and sections of varying roughness, polystyrene layer of sufficient thickness was mounted on a cylindrical column of a wood pattern. The varying diameters and surface roughness for each diameter were carefully generated on the polystyrene using a long-bed lathe machine. This wood-polystyrene unit was then cut longitudinally into two parts. With this half-cylinder forming a core, the concrete half-block was formed by pouring the special concrete mixture into an already prepared large wooden pattern. After proper curing of the concrete, the wooden pattern was disassembled, the polystyrene core was completely removed, and the block was carefully cleaned. The process is summarized in **Figures 3** and **4**. A close-up view showing the variation of diameter and roughness in the concrete half-block are presented in **Figure 5**. The inside and outside packers were also precision-cut into two longitudinal halves as shown in **Figures 6** and 7. The tubulars and packers were carefully positioned and fitted inside the concrete half-block, and the assembly was then fitted with plexi-glass front and top walls. A hook-type lifting mechanism was fixed onto the top (**Figures 8** and **9**) to move the full assembly from one place to the other, using overhead crane or other moving/lifting equipment.

The most challenging design task of this experimental unit was to make it leak-proof. Although many common techniques were available, it had to be guaranteed that it will remain leak-proof even after substantial swelling of elastomers, which can exert enormous pressures. A detailed comparative study was carried out about all available varieties of weather-shield paints, and their merits and demerits, but all had to be discarded due to pressure sustainability requirements. Finally, a carefully selected epoxy coating was smeared onto the walls before the plexi-glass sheets were bolted into position. The torque applied on to the bolts fastening the

Figure 3.
Concreting of the half-block with polystyrene half-cylinder carefully placed inside; and curing of the assembly.

Figure 4.
Removing of polystyrene core (left); final shape of the setup with varying roughness and hole diameter (right).

Figure 5.
Close-up showing variation of diameter and roughness in the concrete half-block.

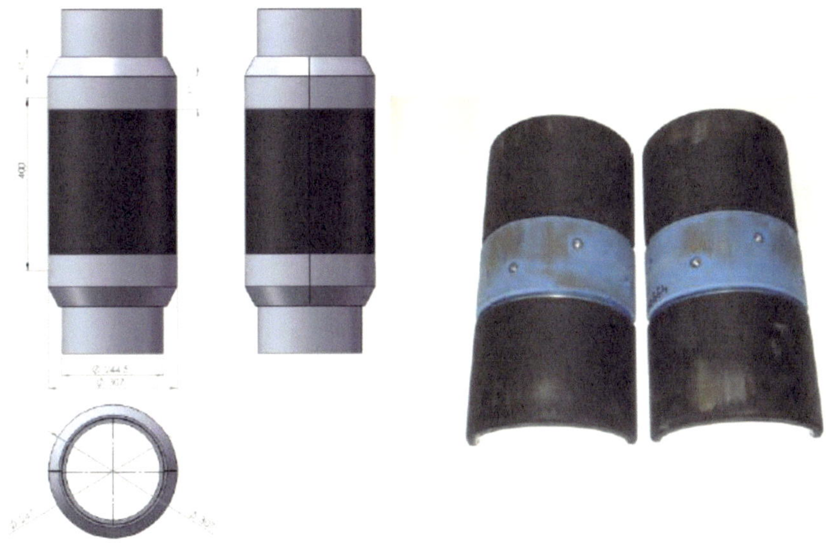

Figure 6.
Dimensions and cut-away sections of the swell packer.

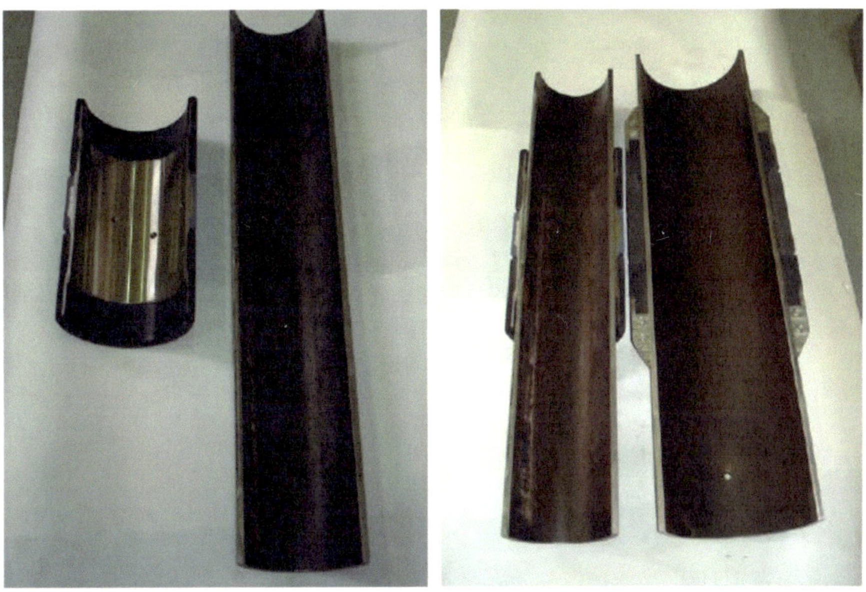

Figure 7.
Half portions of inside swell packer and pipe (left); inside and outside swell packers fitted on respective pipes (right).

plexi-glass and concrete block were calculated using standard procedure outlined in design of fasteners [14]. Similarly, to make it water proof, a combination of Tank Guard 1 and 2 paints from Jotun Marine Paints Division was used so that the concrete half-block could remain water-proof even when submerged with water from the inside for years of use.

At the end, designed-for-purpose screw was removed from the top, unit was filled up with water through a small funnel, and the screws were retightened. An automatic camera was set up on a tripod, focusing on the swelling elastomer region to periodically

Figure 8.
Fitting of the tubular in the concrete block, together with the lifting arrangement (left); fixing of the plexi-glass front wall (right).

Figure 9.
Assembled setup with inner and outer packers, plexi-glass front and top walls, and lifting arrangement.

record the swelling process using its advanced time-step programming module. Manual photographic and video recording was also done at fixed time intervals.

4. Results and discussion

After observing and monitoring the swelling behavior of the packers for a few days, it was found that the steel tubulars and screws/bolts exposed to water had started to corrode. This would have only a minor effect on swelling, but the rust particles diffused into the water and visibility reduced drastically after more corrosion took place as shown in **Figure 10**. To remedy this problem, the water was siphoned off using a portable pump, and the rust stains and particles were vigorously cleaned. This flushing-cleaning of the unit improved the visibility significantly (**Figure 11**). Under actual shallow-aquifer field conditions of around 50 degrees Celsius temperature and 0.5% salinity, 70% swelling had been predicted for the outer packer (elastomer E1) in six hours during swelling tests conducted on laboratory size samples [9, 10], confirming its fast swell character. Fresh water used in test unit behaves almost the same in terms of salinity as the 0.5% brine solution, so this field condition was almost replicated. However, the unit being kept in an air-conditioned lab meant that the test temperature was nearly 20 degrees Celsius, a large difference from the field temperature, which definitely reduced the swelling rate of the elastomer by a significant percent. Also, the packer in the test unit could swell only in the outward direction, and was swelling against a concrete hole of varying diameter and roughness, markedly different from lab-size elastomer samples swelling freely on all sides. It was thus no surprise that it took more than 10 days for the elastomer to fill the irregular cavities in the concrete half-block.

The inside swell packer (E2) was intended for an application with salinity ranging from 12–18%, and working temperature environment of 40 degrees to 90 degrees Celsius environments. This test unit could be classified as a low-salinity and medium-temperature setting. According to earlier lab tests [9, 10], it was estimated to reach 70% swelling in 3 to 4 weeks. This was not too far off as it took the inner swell packer almost 5 weeks to swell against the casing, even though the salinity and temperature conditions were different. There was apprehension about the behavior of this elastomer under fresh water. However, it remained glued firmly to the tubular. **Figures 12** and **13** give an idea of the behavior of fast-swell and inside swell packers before exposure to water, and at different stages after being in contact with water.

There were intermittent leakages from the sides or the bottom, particularly at the interface between plexi-glass and the concrete block or casing or production

Figure 10.
Assembled unit after initial water fill-up (left); rust stains after 8 days of water exposure (right).

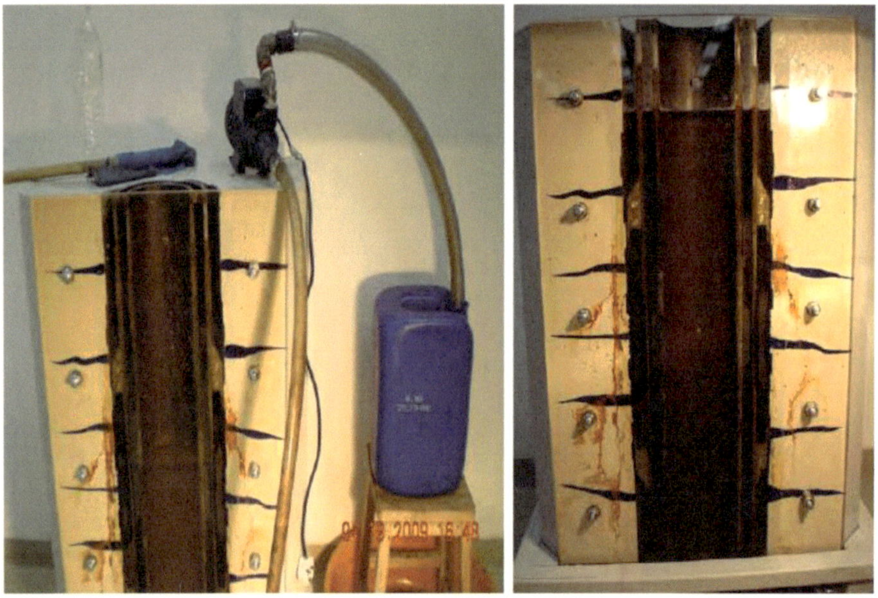

Figure 11.
Flushing and cleaning of the unit (left); water fill-up after flushing (right).

Figure 12.
Elastomer seals before exposure to water (left), and after near-complete swelling of the outer packer (right); inner packer has only partially expanded.

tubing. Most of these leakages were completely stopped by applying more torque on the bolts, thus increasing the tightness between the front-end plexi-glass to the concrete block. Also, tubulars and bolts started to rust again after a few weeks, reducing the visibility once more. Periodic flushing-cleaning would not be a good solution if the unit were to be used for long-term duration. It was thus decided to dismantle the entire system for a major unit overhaul.

Figure 13.
Near-complete swelling of the outer elastomer seals against cavities of varying diameter and roughness.

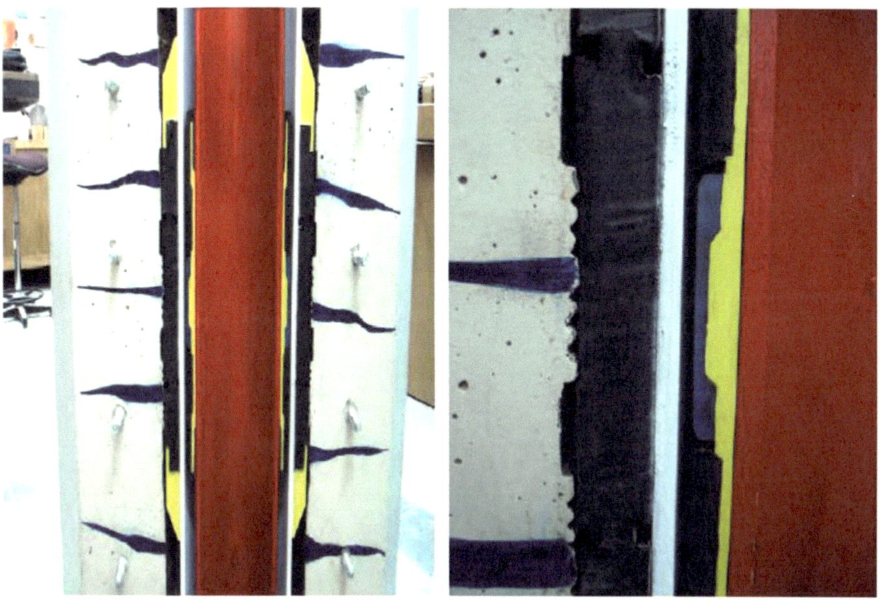

Figure 14.
Demonstration-test unit after disassembly, flushing and cleaning, repainting of different components, and reassembly; partial de-swelling can be observed.

Before disassembling the unit, research was again conducted about the most suitable water-shield paints for steels. Water was drained out from the unit, and it was moved back to the laboratory as quickly as possible, to maintain the same environment as before in order to avoid significant de-swelling of the elastomers. The complete unit was disassembled and thoroughly cleaned using wipes and different grades of sand and emery paper. Some steel bolts were replaced by non-corroding brass ones. Insides

Figure 15.
Re-swelling of the elastomer seals after disassembly, reassembly and water re-fill.

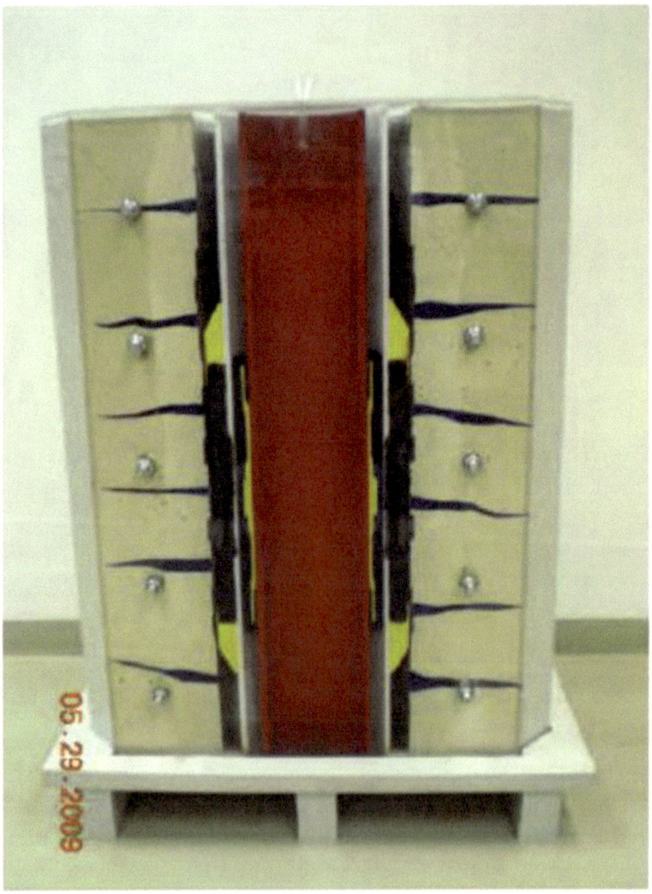

Figure 16.
Experimental unit after one year of re-swell showing the swelling durability.

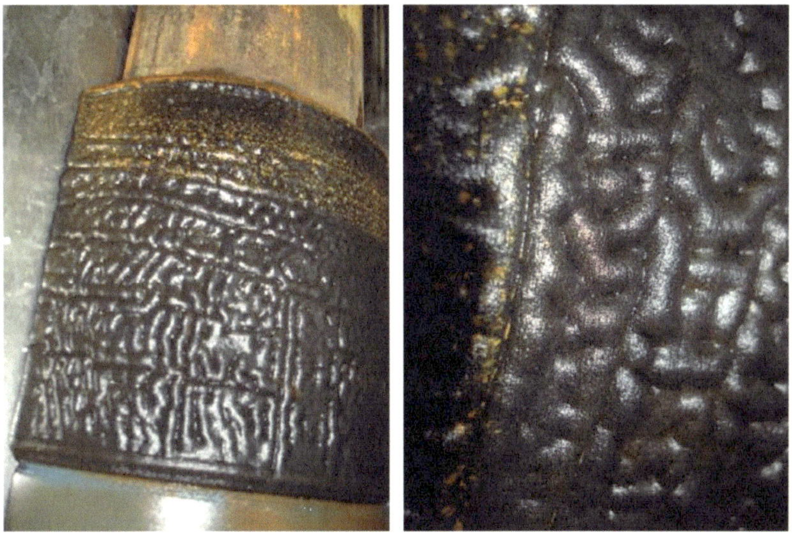

Figure 17.
Pictures of the elastomer surface that swells against formation (concrete block), after swelling and partial de-swelling.

of the concrete block were repainted with appropriate layers of tank-guard protection. Tubulars and exposed bolts were painted with red oxide primer and weather-shield coatings of different colors, for rust-protection as well as improved identification of different parts. Fresh epoxy coating was applied to the concrete front-wall before fixing the plexi-glass sheets. However, with all such measures, a small amount of de-swelling had taken place during the overhaul process of the experimental unit.

Once assembled and filled again with water, the two swell packers started to swell again as shown in **Figures 14** and **15**. Long-term contact of packers and concrete and metal elements to standing water without any flushing could pose another serious problem. Algae and other biological growth can crop up in the water over time. This can have unseen effects on the swelling setup, and can again reduce the visibility to very low levels. In order to alleviate this problem, a chemical was identified that inhibits bacterial and other biological growth in standing water. When water was filled into the reassembled unit, this chemical was added to the water in measured quantity. With the help of these modifications/improvements, the system exhibited improved visibility and no leakage even after several months of continuous water exposure as shown in **Figure 16**.

Dismantling the system gave a great unexpected benefit. Generally, no one has a clear idea of how the elastomer looks after swelling against the casing or the formation. However, as the system was disassembled, there was a chance to have a close look at the elastomer surface that was swelling against the concrete block. **Figure 17** displays extremely interesting and unique photographs of the almost brain-like elastomer surface that is pressed against the concrete block due to swelling. Further investigation of this behavior could reveal interesting physical phenomena responsible for this pattern after swelling and partial de-swelling of elastomers in general, and of these specific types of elastomers in particular.

5. Conclusions

An experimental-cum-demonstration unit was designed, fabricated and assembled for testing of swelling elastomer seals against a casing (cased hole), and against the rock formation (open hole). A concrete half-block with a semicircular hole was designed and fabricated with varying inside diameters and surface roughness to reproduce the effect of actual formation, with front and top walls made of plexiglass for clear visibility. Smaller packer on 177.8 mm inner diameter steel tubular had to swell against a 244.475 mm casing, while the larger diameter swell packer (fast swell elastomer) on the casing had to swell on 311.15 diameter of formation represented by the concrete block. Given the difference in field and test conditions (0.5% salinity water at 50 degrees Celsius, against fresh water at 20 degrees' Celsius ambient temperature), 70% swelling results were very reasonable. The fast swell packer took about 10 days, while the other swell packer took around 5 weeks to completely fill all irregular surfaces. Modifications and improvements were carried out on the unit to make it reliably water-proof and leak-proof, and to improve and maintain the visibility. This experimental unit also a one of its kind demonstration unit, reproducing the behavior of swell packers in cased and open holes, and providing visual confirmation to the field engineers of the swelling of elastomer seals against casing and formation.

Author details

Sayyad Zahid Qamar[1*], Maaz Akhtar[2] and Tasneem Pervez[1]

1 Mechanical and Industrial Engineering Department, Sultan Qaboos University, Muscat, Oman

2 Mechanical Engineering Department, N.E.D. University of Engineering and Technology, Karachi, Pakistan

*Address all correspondence to: sayyad@squ.edu.om

References

[1] Oilfield Glossary (2019) "Packer," Schlumberger, accessed July 2019, https://www.glossary.oilfield.slb.com/en/Terms/p/packer.aspx

[2] PetroWiki (2019) "Packer," https://petrowiki.org/Packers, accessed July 2019

[3] Triolo MT, Anderson LF, Smith MV (2002) "Resolving the Completion Engineer's Dilemma: Permanent or Retrievable Packer?" Paper # SPE-76711, SPE Western Regional/AAPG Pacific Section Joint Meeting, 20-22 May 2002, Anchorage, Alaska, USA

[4] Bell WT, Buzarde LE, DePriester CL, Kastor RL (1972) "Production Operations Course 1 - Well Completions," Dallas Texas 75206: Society of Petroleum Engineers AIME.

[5] Wikipedia (2019) "Production Packer," https://en.wikipedia.org/wiki/Production_packer, accessed July 2019.

[6] Dorohov M, Kostryba I, Biletskyi V (2016) "Experimental Research on Sealing Ability of Borehole Packers," Eastern European Journal of Enterprise Technologies, 82 (4/8), p 56-68

[7] Rigzone (2019) "How Does a Swellable Packer Work?" https://www.rigzone.com/training/insight.asp?insight_id=353&c_id, accessed July 2019.

[8] Qamar SZ, Pervez T, Akhtar M, Al-Kharusi MSM (2012) "Design and Manufacture of Swell Packers: Influence of Material Behavior," Journal of Materials and Manufacturing Processes, 27 (7), p 721-726, 2012.

[9] Qamar SZ, Hiddabi SA, Pervez T, Marketz F (2009) "Mechanical Testing and Characterization of a Swelling Elastomer," Journal of Elastomers and Plastics, 41 (5), September 2009, p 415-431

[10] Qamar SZ, Pervez T, Akhtar M (2016) "Performance Evaluation of Water-Swelling and Oil-Swelling Elastomers," Journal of Elastomers and Plastics, 48 (6) 2016, p 535-545

[11] Al-Yami A, Nasr-El-Din H, Al-Saleh S, Al-Humaidi A, Al-Arfaj M, Awang M, Al-Mohanna K (2008) "Lab Investigation of Oil Swelling Elastomers for Smart Well," Paper # SPE 19403, SPE Offshore Technology Conference, Houston, Texas, USA

[12] Patel H, Salehi S, Ahmed R, Teodoriu C (2019) "Review of Elastomer Seal Assemblies in Oil and Gas Wells: Performance Evaluation, Failure Mechanisms, and Gaps in Industry Standards," Journal of Petroleum Science and Engineering, 179 (2019), p 1046-1062

[13] Pervez T, Qamar SZQ (2011) "Full Scale Testing of Fast Swell Packer for Shallow Aquifer Applications," Paper # SPE/IADC 148533, SPE/IADC Middle East Drilling Technology Conference, 24-26 October 2011, Muscat, Oman

[14] Norton RL (2012) Machine Design: An Integrated Approach, 5th edition, Pearson Education.

Chapter 5

Long-Term Integrity Testing of Water-Swelling and Oil-Swelling Packers

Sayyad Zahid Qamar, Maaz Akhtar and Tasneem Pervez

An experiment is never a failure solely because it fails to achieve predicted results. An experiment is a failure only when it also fails adequately to test the hypothesis in question, when the data it produces do not prove anything one way or another.

Robert Pirsig

Abstract

As easy oil in many fields is dwindling, there is increasing stress worldwide on innovative enhanced oil recovery (EOR) techniques. One forward-looking EOR approach is the workover method. It tries to convert currently weak horizontal wells to maximum reservoir contact (MRC) wells, or abandoned vertical wells to horizontal ones or power water injectors. Where conventional techniques fail, swelling elastomer seals and packers provide effective water shutoff and zonal isolation in even very complex environments, resulting in significant savings in rig time and development cost. One major issue of interest is the service life of elastomer seals and packers. It can be attempted to predict the probable working life based on the theory of accelerated testing. However, this forecast will not be very dependable for swelling elastomers as the material performance is substantially different from other rubber-type polymers. A full-scale test rig (one of its kind in the world) was therefore designed and fabricated at Sultan Qaboos University (SQU), in collaboration with a regional petroleum development company, for long-term service life assessment of actual full-size water-swelling and oil-swelling packers.

Keywords: Swell packers, in-situ longevity testing, water-swelling, oil-swelling

1. Introduction

As easy oil in many fields is dwindling, there is increasing stress worldwide on innovative enhanced oil recovery (EOR) techniques [1–6]. One forward-looking EOR approach is the workover method. It tries to convert currently weak horizontal wells to maximum reservoir contact (MRC) wells, or abandoned vertical wells to horizontal ones or power water injectors [7, 8]. This is done to maximize well production and to achieve total oil recovery. This requires the placement of smart systems in the well for controlling of flow from each lateral. This type of zonal isolation is built around expandable liners and swell packers.

Intelligent and multilateral wells form another strategy for maximum hydrocarbon recovery. Efficient zonal isolation and reservoir compartmentalization are the keys to success for these well systems. Where conventional techniques fail, swelling elastomer seals and packers provide effective water shutoff and zonal isolation in even very complex environments, resulting in significant savings in rig time and development cost [9]. One major issue of interest is the service life of elastomer seals and packers [10, 11]. As yet, no data is available from service providers, designers, or manufacturers about the durability or long-term endurance of swell packers under actual well conditions. It can be attempted to predict the probable working life based on the theory of accelerated testing [12]. However, this forecast will not be very dependable for swelling elastomers as the material performance is substantially different from other rubber-type polymers. A full-scale test rig (one of its kind in the world) was therefore designed and fabricated at Sultan Qaboos University (SQU), in collaboration with a regional petroleum development company, for long-term service life assessment of actual full-size water-swelling and oil-swelling packers [13].

These packers are of different elastomer materials, kept in crude oil and saline solutions at different temperatures, and exposed to high pressures. The design process went through the typical stages of specifications development, concept design and evaluation, detail design, and assessment for reliability and manufacturability. The test setup was built around some important modules: thermal system with capability of maintaining elevated temperatures continuously over a 5-year period; recirculation system to keep salinity at the requisite level; arrangement to pressurize the packers after sealing has been achieved through elastomer swelling, and maintaining it for several years; and a complex system for temperature and pressure measurement on both upstream and downstream sections in all packers. This unique long-term reliability assessment study is expected to provide helpful pointers to field engineers and application designers in appropriate selection of swell packers and in packer design enhancement.

2. Specifications

Published literature and information from vendor websites were critically reviewed. Focus was on works related to deployment of water-swelling and oil-swelling elastomers in oil and gas wells, new development and remediation efforts, and relevant well conditions (salt concentration of brine, type of crude oil, and in-situ temperature and pressure). After a series of discussions with local and regional field engineers, proposal for a test setup for long-term durability assessment of swell packers was agreed upon. Following were the key specifications. There will be ten units, including nine actual packers placed inside actual steel casings (as in real wells). One display unit will have a transparent Perspex outer pipe, so that internal details could be clearly seen which are not visible in the actual units because of the outer steel casings. Outer casings will be real 7-in steel tubulars. Packer nominal diameter will be 3½-in for four tubes, and 4½-in for the other five. Six units will have water-swelling elastomers, and 3 units will be of the oil-swelling type. Salt water solutions will be of two concentrations: 0.5% in two units, as in low-salinity wells; and 12% in four tubes, to represent high-salinity wells. Oil-swell packers will have actual crude oil from two different regional wells. Two test temperatures will be maintained: room temperature to typify shallow-aquifer type wells; and 50°C representing medium-depth wells. After sealing, four packers will be subjected to a pressure of 1000 psi, characteristic pressure range in many of the well types being studied. Tests will continue for a 5-year duration.

3. Test rig design

Several feasible concepts were developed, followed by thorough design evaluation. Significant features of the selected design are described in this section. A circulation system was needed to maintain water salinity at required levels; without it, salt would precipitate out and salinity would go down. Its main components were water heaters and containers, circulation pumps with control units, and circulation pipes. A thermal system was necessary to continually heat selected packers to 50°C for five years. It consisted of thermal blankets, insulation, control system, and temperature gauges. A pressurizing unit was needed to apply and maintain high pressure in the packers after sealing was achieved through swelling. Its critical components were a pressure-manifold (connected to all packers), high-pressure source, high-pressure pipes and connections, and pressure gauges. A monitoring system was required to observe temperatures and pressures in all units, between the packer and the casing, and inside the inner tube. A detection system had to be there to signal the completion of sealing in each packer, achieved by swelling of the elastomer against the outer casing. A second detection system was needed to indicate any seal failure (after seal completion). A sturdy frame was required as a support and housing for all the test and demo units. Concept evaluation was done using Pugh's basic decision matrix method. Overriding criteria were safety, reliability, and a minimum of 5-year service life. All components (including valves, fittings, and welded joints) were required to have a minimum rating of 100 bar pressure, to provide a safety blanket for the design pressure of 70 bar (about 1000 psi).

Figure 1 schematically shows the arrangement of the different test units. An identification number is assigned to each unit, and brief descriptions are

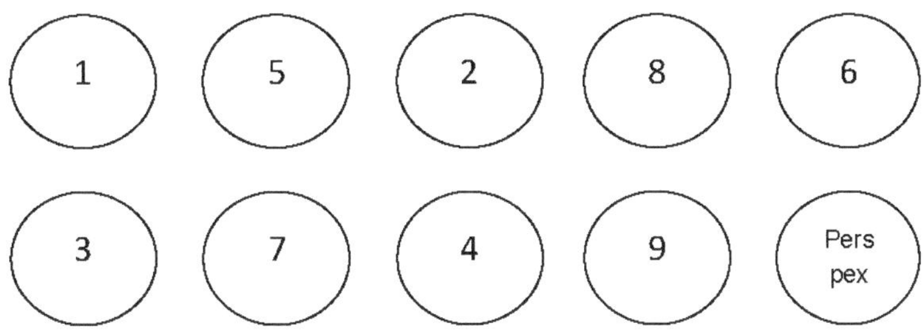

Figure 1.
Layout of the longevity test setup.

Unit #	Elastomer Type	Swelling Medium	Temperature	Unit #	Elastomer Type	Swelling Medium	Temperature
	3½-in swell packer inside 7-in casing				**4½-in swell packer inside 7-in casing**		
1	W2	12% brine	Room temp	5	W2	12% brine	Room temp
2	O1	Crude oil	Room temp	6	O1	Crude oil	Room temp
3	W2	12% brine	50°C	7	W2	12% brine	50°C
4	O1	Crude oil	50°C	8	W1	0.5% brine	Room temp
	Perspex demonstration unit			9	W1	0.5% brine	50°C
10	W1	—	Room temp				

Table 1.
Elements of the longevity test setup.

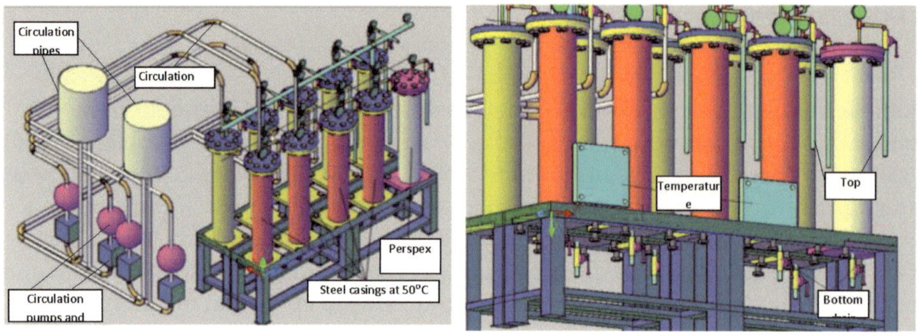

Figure 2.
Schematic diagram of the test facility showing test and demonstration units; circulation system (tanks, pipes, pumps, and controllers); thermal system and controllers; top and bottom drain systems.

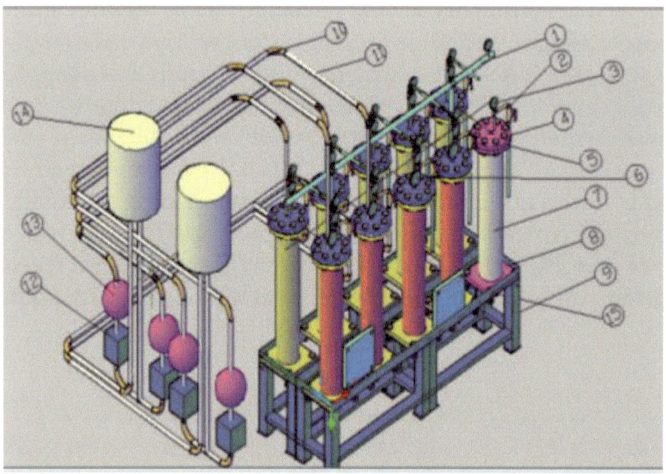

Figure 3.
Schematic assembly drawing of the longevity test setup. 1. Pressure manifold; 2. Drain system; 3. Pressure gauges; 4. Nuts and bolts; 5. Flanges; 6. Temperature gauges; 7. Casing tubulars; 8. Bottom plates; 9. Base frame; 10. Elbow joints; 11. Circulation pipes; 12. Circulation pump controllers; 13. Circulation pumps; 14. Water heaters/tanks; 15. Temperature controllers; 16. High pressure system; 17. Perspex demonstration unit.

summarized in **Table 1**. **Figures 2** and **3** are schematic drawings describing configuration and layout of all components of the test rig.

4. Fabrication and assembly

Swelling elastomer pipes and outer steel casings were provided by a regional petroleum development company. Elastomer sections were mounted on steel tubulars, different ones for low-salinity and high-salinity water, and crude oil; **Figure 4**. Sections of one-meter length were cut out from these tubulars, followed by sizing and trimming to get the swell packers and outer casings. Beveling was required for welding operations to be carried out at one end of the packers, and both ends on the casings; **Figure 4**.

Flanges, blind flanges, and bottom end plates were fabricated for the packers and the casings. Pitch circle bolt-holes (8 in number) had to be drilled in the flanges for joining the casings to the packers. Drilling of corner holes (4 in number) in the bottom plates was needed to fasten the casings to the base frame. Threaded holes

Figure 4.
Full-length elastomer tubulars (left); sizing and beveling of elastomer packers (right).

Figure 5.
Fabrication of flanges (left); welding of flanges and bottom plates to packers and casings.

Figure 6.
Fixing of drain valves to bottom plates (left); base frame to hold the test units (right).

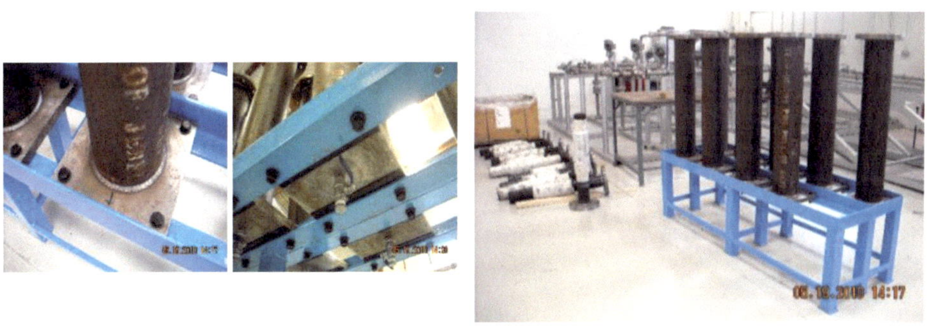

Figure 7.
Fixing of casings to base frame (left); placement of packers inside casings (right).

(6 in number) with proper spacing were drilled in the top flanges to house the pressure and temperature gauges; **Figure 5**. Welding of top flanges and bottom plates to packers and casings completed this step of the assembly; **Figure 5**.

Each bottom plate had a single hole drilled in the center, to attach the drain valve. Drainage lines were fixed at the bottom of each unit, with stainless steel (SS)

Figure 8.
Thermal system components (left); winding of thermal blankets on casings (right).

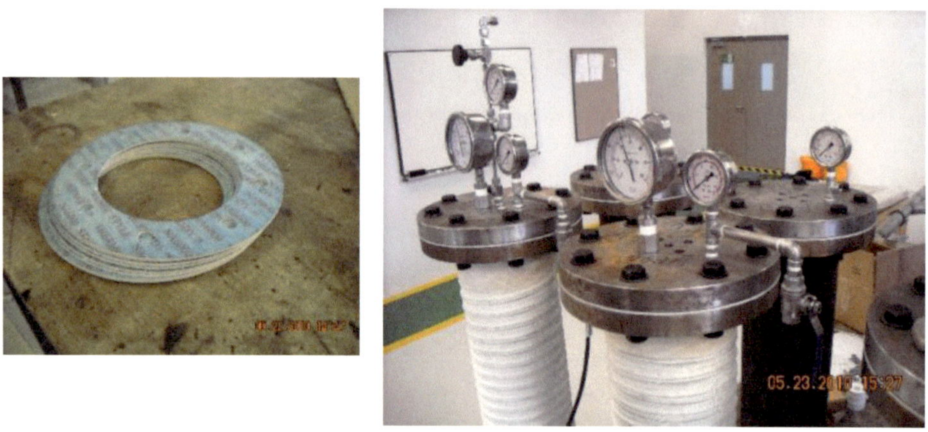

Figure 9.
High temperature–pressure gaskets (left); joining of packer and casing flanges, and fixing of pressure and temperature gauges (right).

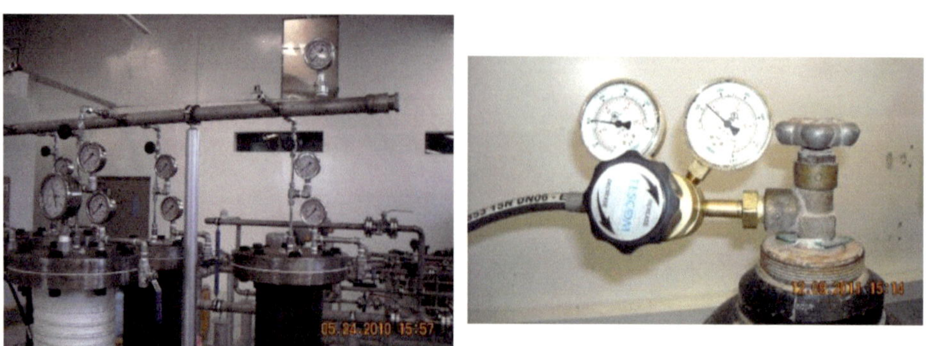

Figure 10.
High-pressure manifold with connections to test un its (left); pressure regulator to control high pressure flow into the test units (right).

ball valves and plugs; **Figure 6**. Upper drainage lines were also fitted through the top flange of each of the ten units, together with SS high-pressure (100-bar) ball valves and pressure gauges. A base-frame was fabricated from steel channels etc. to provide housing and support to all packer units; **Figure 6**. Casings were attached to the base frame using bolts, and packers were fitted inside the outer casings; **Figure 7**.

Thermal blanket system was imported from a company that specialized in custom-designed heating units. Assembly was done in-house and the thermal units were installed on four high-temperature packers; **Figure 8**. Joining of packers to casings was done flange-to-flange, with custom-built pressure seals in between. Pressure gauges were fitted on top of the upper flanges for all units, and temperature gauges for the four 50°C units; **Figure 9**.

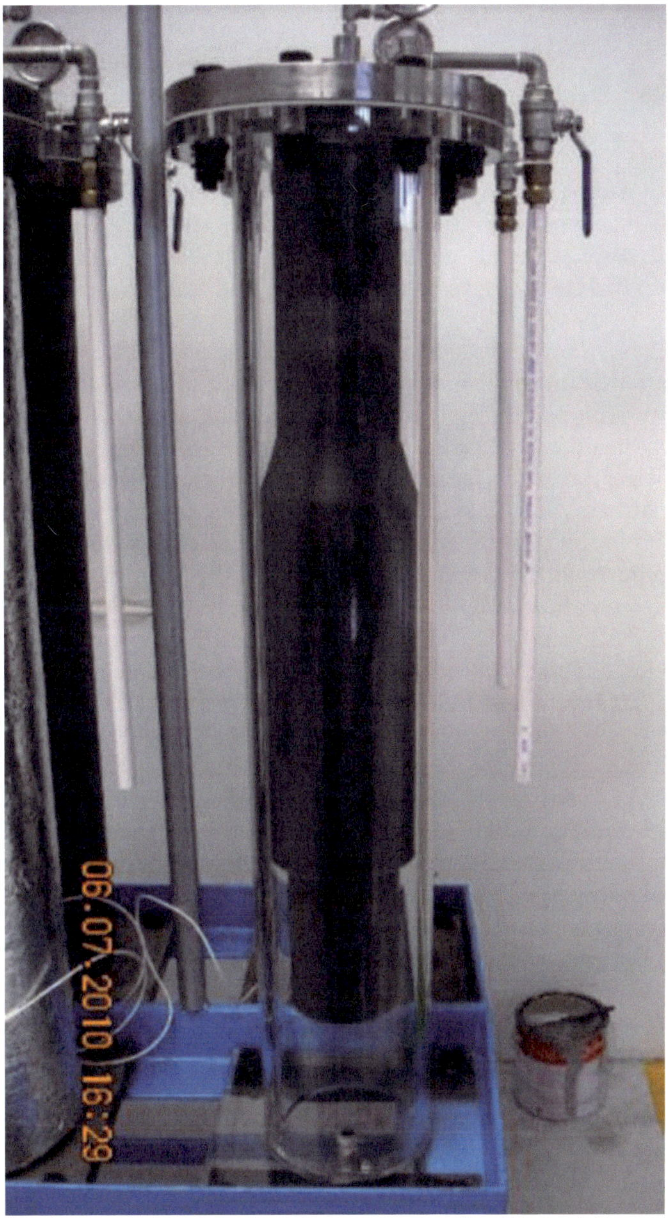

Figure 11.
Perspex demonstration unit showing internal construction of other test units.

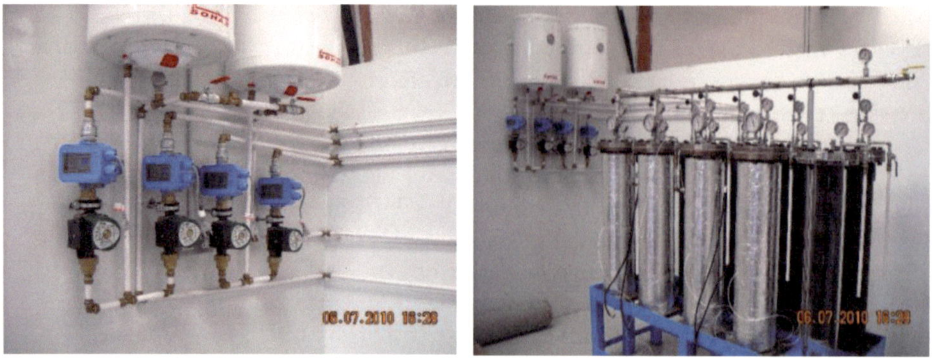

Figure 12.
Ambient and hot water circulation system (left); complete assembly of the longevity test setup.

A manifold system was fabricated and installed, using special valves, to apply and maintain high pressure when needed; **Figure 10**. Main part of this manifold system was a SS pipe that was to be later connected to a high-pressure nitrogen gas cylinder. It had nine specially-fabricated outlets, hooked up to the test units, containing SS high-pressure (100-bar) needle valves and pressure gauges. High-pressure multistage gas regulator and related fittings were affixed to the nitrogen cylinder, to control the flow pressure in the manifold and the packer units; **Figure 10**.

Detail and assembly drawings for the demonstration unit were sent to a specialist facility for fabrication, the outer casing and the blind flange to be made of transparent Perspex. This see-through unit had all the components and was assembled to exactly match the other functional test units; **Figure 11**.

A circulation system was fabricated and installed on the four water-swelling packers, to maintain the salt concentration. It consisted of two water heaters (50-ltrs), three circulating pumps, and four each of pressure controllers, non-return valves, filling ports with valves, air vents, and running-time meters. All of these were connected together through copper lines; **Figure 12**.

Dedicated units were installed on all nine packers to carry out salt-water and crude-oil filling, and to check whether swell packers had sealed against the outer casing or not. This system included special SS ball valves at inlet and outlet ports.

All the components and sub-assemblies were finally assembled together; **Figure 12**. Cutting, trimming, beveling, drilling, threading, bending, and welding were some of the major fabrication operations. SS and PVC pipes, flexible hoses, valves, pressure and temperature gages, circulation and heating system, etc. were fitted at required locations. Forty sets of M16 x 100 and 80 sets of M16 x 80 hexagonal bolts (together with nuts, washers, and spring washers) were employed for different types of connections: packer blind flanges to flanges on outer casings; bottom end plates on casings to base frame; etc.

5. Commissioning and preliminary testing

Various stages of the fabrication, construction, subassembly and assembly work were described above. After that, pilot tests were run for working inspection of the water circulation system, the thermal blanket system, and the high-pressure manifold system. After several rounds of these trial runs, the longevity test rig was ready for commissioning.

Approximately 25-ltr of saline water was required to fill out each packer unit: complete filling inside the swell packer, and then up to a requisite height above the elastomer section in the annular space between the packer and the outer casing. This necessitated the preparation of a very large amount of distilled water: 25 liters per unit for six water-swelling units. This was done with the help of the water purification setup in the Environment Lab of the College of Engineering. Using this distilled water, salt solutions of 0.5% and 12% salinity were prepared, in enough quantity to be stored in 25-ltr storage canisters for initial filling of the tubes, and for intermittent later re-fillings due to possible leakages etc. Proper packaging and transportation to the test rig of a large quantity of crude oil (25-ltr for each of the three oil-swelling units) from two local oilfields was also a formidable task. An important requirement was that testing had to begin simultaneously for all water-swelling and oil-swelling packers, to have the same swelling time. Full commissioning and testing were the next steps.

6. Initial problems

Certain unique problems were encountered in the first couple of weeks of commissioning of the test facility, and were successfully resolved, before the system could be declared fully functional.

6.1 Water filling

Four tubes were to be filled with 12% brine solution, and two units with 0.5% saline water. Manual filling was initially attempted, but proved to be very time consuming. In case the filling took a few days, elastomer seals in some tubes would swell and seal before the other units. Suitable pumps were therefore utilized to accelerate the process and achieve simultaneous filling of all water-swell units.

6.2 Oil filling

It was almost impossible to do oil filling manually, as the viscosity of the crude oil was too high to easily pass through the filling ports under normal pressure. Use of regular water-type pumps could not work easily for pumping of such extra-thick crude oil. It was required to repeatedly re-prime and re-fill the pumps throughout the operation. Not only was the filling process slowed down, but three pumps were burnt out before oil filling could be completed.

6.3 Circulation system problems

As mentioned above, a circulation system was needed to maintain salt-concentration in the 4 high-salinity (12%) units. Small leakages repeatedly occurred in various portions of the circulation system, causing salt deposits on pipes and fittings. Water leakage and salt deposition also caused rusting of the tubes, flanges, and other fittings; **Figure 13**. The system had to be monitored on a daily basis, and clean-up, re-tightening and re-sealing operations had to be frequently carried out. Salt leakage also meant that water salinity would change. So salinity checks and refilling of tubes was also needed occasionally.

Intermittent shutting down of circulation pumps was another problem, caused by the formation of air gaps in the system due to small leakages. Urgent refilling and bleeding were needed to solve this problem, to avoid any notable downtime in the

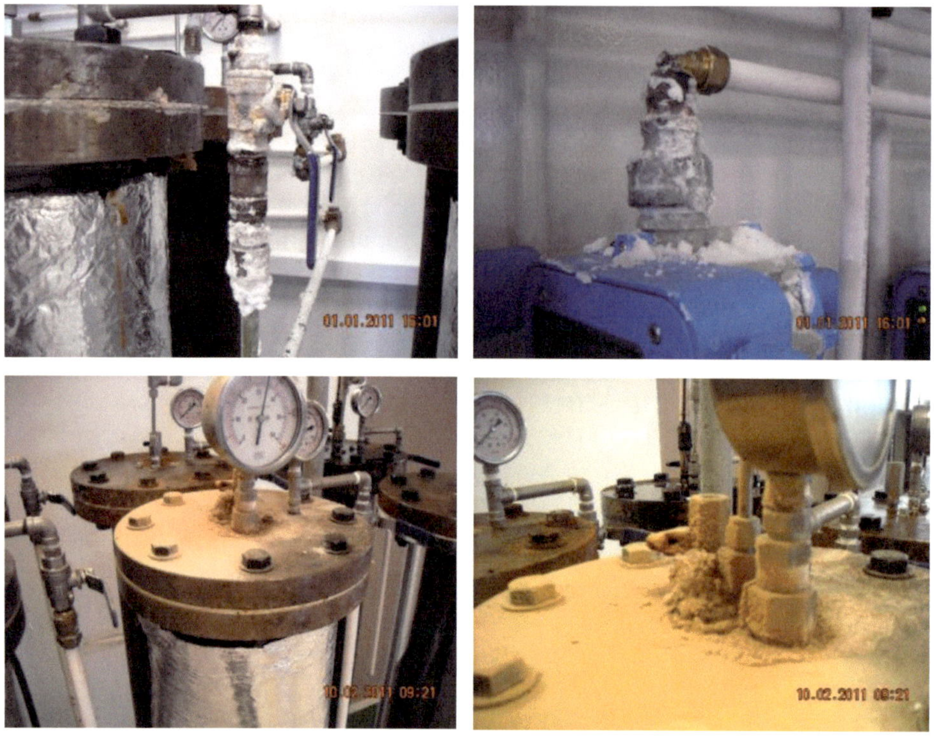

Figure 13.
Salt deposition on pipes, fittings, and pump units due to water leakage (top); corrosion of pipes and fasteners due to water leakage and salt deposition (bottom).

circulation system; otherwise salinity levels could not be maintained. Careful daily monitoring of the test rig was therefore kept up.

6.4 Maintaining temperature

The thermal blanket system was fitted with automatic control units to keep temperature in hot tubes at a constant value of 50°C. However, because of small thermal leakages to the base frame and the surroundings, temperature in some tubes dropped a little from time to time. Close scrutiny of the temperature gauges was required (on a daily basis), together with minor adjustments of the temperature-setting dial.

7. Testing and monitoring

Detailed log of temperature and pressure readings on all tubes, and of any uncommon occurrences was maintained throughout the five-year test period. This was done on a daily basis initially, then every two days, and then once per week.

7.1 Seal check

Initially, daily checks were carried out to see if any tube had sealed. Based on previous experience of material-testing for a variety of swelling elastomers, earlier sealing was expected in low-salinity water, then in high-salinity water, and then in oil. Also, faster sealing was expected for the higher temperature units [10, 11]. After some time, when quick sealing was not observed, the seal-check duration was changed to two days. Circulation pump would automatically stop for a tube once it was sealed; seal check was therefore carried out every time a pump stopped. When it was found that sealing was not complete, and the pump had stopped only because of an air gap, refilling and bleeding were carried out to re-start the pump.

7.2 High-pressure testing

The initial plan was to pressurize all four high-pressure tubes to 1000 psi simultaneously. However, it was observed that some of the tubes did not seal even after a few months. A decision was thus taken to pressurize each of these tubes as they sealed. This required connecting a high-pressure nitrogen cylinder to the pressure-manifold. Seal integrity was re-checked for the tube to be pressurized. If found intact, the high-pressure inlet valve on the manifold for this tube was opened and pressure was carefully increased in steps of 10 bar, as a safeguard against possible sudden seal failure. After reaching the full pressure of 70 bar, the tube was observed for about 20 min before the manifold valve was closed and the unit was disconnected from the nitrogen source. Pressure maintenance in this unit was guaranteed by the one-way valve. A drop in pressure of 10 to 20 bar was observed in the pressurized packers over the next few days. Re-pressurization to 1000 psi was then carried out. Absorption of some nitrogen into the salt solution (or into the elastomer material) may have caused this pressure reduction.

8. Results, observations and discussion

Some notable observations during the first few months, and through the 5-year test period, are mentioned below. It should be noted that 'W1' is a low-salinity fast-swelling water-based elastomer, 'W2' is a high-salinity medium-swelling water-based elastomer, and 'O1' is an oil-based elastomer. The gap between elastomer and the 7-in outer-casing is larger for the 3½-inch packer as compared to the 4½-inch packer; elastomer would have to swell an extra half-inch for sealing to be completed in the 3½-inch units.

8.1 Sealing time

Sealing times, in sequence, for the nine test tubes (swell packers) are summarized in **Table 2**. Of special interest is tube-5 (W2 elastomer; 4½-inch packer; 12% salt solution; room temperature) which sealed in 134 days, de-sealed after roughly 6 months, and re-sealed sometime after draining and re-filling of the brine solution.

8.1.1 Discussion

Let us recall some major findings from earlier studies [14–17] on swelling elastomer testing and characterization, and performance evaluation of elastomer seals and packers, conducted by the authors. Elastomers swell more in low-salinity brine than in salt solution of higher salinity. Water-swelling elastomers swell more and at a faster rate than oil-swelling elastomers. Higher amount of swelling takes place at higher temperatures. Elastomers developed for lower salinity may not perform well in higher salinity environment. If packers were stacked in open yards for long time, their performance seriously went down due to exposure in comparison with fresh packers; smaller amount of total swelling, and at a much slower rate.

In view of these earlier experimental conclusions, most of the observations listed in **Table 2** are as expected, and have rational explanations. Tube-9;

Tube Number	Sealing Time (Days)
9	8
8	43
3	62
7	115
5	Sealed in 134 days; de-sealed after 6 months; re-sealed after some time
4	166
1	178
2	206
6	Did not seal

Table 2.
Sealing times for the nine test tubes (swell packers).

fast-swelling elastomer (W1), low-salinity brine (0.5%), small elastomer-casing gap (4½-inch packer), and higher-temperature (50°C). Naturally, it was the first to seal (8 days). Tube-8; fast-swelling elastomer (W1), low-salinity brine (0.5%), small gap (4½-inch packer), room-temperature; second in seal completion (43 days), as expected. Tube-3; medium-swelling elastomer (W2), low-salinity brine (0.5%), larger gap (3½-inch packer), higher-temperature (50°C); third to seal (62 days). Though gap was larger, low-salinity and high-temperature combined to yield faster swelling. Tube-7; medium-swelling elastomer (W2), high-salinity brine (12%), small gap (4½-inch packer), higher-temperature (50°C); fourth to seal (115 days). While the gap was smaller than tube-3, much higher salinity (12% compared to 0.5%) resulted in slower swelling rate. Tube-5; medium-swelling elastomer (W2), high-salinity brine (12%), small gap (4½-inch packer), room-temperature; fifth in sealing (134 days). As expected; slower swelling rate for combination of higher salinity and lower temperature. Tube-4; oil-swelling elastomer (O1), crude oil, large gap (3½-inch packer), high temperature (50°C); sixth in sealing (166 days). Predictable; oil-swelling elastomer sealed later than both water-swelling elastomers, but before other oil-swelling units due to higher temperature. Tube-1; medium-swelling elastomer (W2), high-salinity brine (12%), large gap (3½-inch packer), room-temperature; seventh in sealing (178 days). Important to note; medium-swelling elastomer seals later than oil-based elastomer if all other conditions are unfavorable (higher salinity, larger gap, lower temperature). Tube-2; oil-swelling elastomer (O1), crude oil, large gap (3½-inch packer), room temperature; eighth to seal (206 days). As expected; longest sealing time for combination of all unfavorable factors: oil-based elastomer, large gap, and lower temperature. Tube-6; oil-swelling elastomer (O1), crude oil, small gap (4½-inch packer), room temperature; did not seal. Unexpected behavior; should have sealed before tube-2; all other parameters were same while gap was smaller. Either elastomer material on this packer segment was sub-standard from the beginning (small manufacturing defects do happen in the best of products), or material degradation due to exposure was much more severe on this segment in comparison with others.

8.1.1.1 Note-1: Long sealing time

One serious anomaly appears to be common in all of the above cases. Except for tube-9 that sealed in a reasonable time (8 days), all other tubes took much longer than expected to swell out and complete the seal. Wells are designed and developed with the target of going into production as fast as possible; production delays translate into significant losses. In an earlier study [17], effect of exposure (to sun and other atmospheric elements) on performance of swelling elastomers was studied. It was found that *performance of packers which were stacked in open yards for several months seriously went down in comparison with fresh packers; smaller amount of swelling, and at a much slower rate*. Packers included in the current longevity study had the same problem; they had been stacked in open yards for very long periods; thus the much slower swelling rate, leading to very long sealing time. This can be the source of a very useful advice to field engineers and procurement personnel. *There should be maximum effort and planning to use almost fresh swell packers. Care must be taken not to order too many packers than actually expected, as old ones will not perform as well. If stacking must be done (to avoid procurement delays), packers should be kept in covered yards, to minimize exposure to the sun and the elements*.

8.1.1.2 Note-2: Multi-segment packer design

In an actual application such as zonal isolation or water-shutoff, there may be water incursion in which water-swellable elastomer would be needed. There can also be oil exposure, needing oil-swelling elastomer. *Alternate segments of water and oil-swell elastomers* are therefore used in practical design [18], as shown in **Figure 14**. Also, if one segment (or more) does not perform satisfactorily due to any material issue, the *series of seal segments guarantees that the region still remains isolated*. Under-performance (no sealing) of tube-6 would not be critical if such multiple-segment design is employed. However, *in the case of single-packer system, operational issues would arise if the packer material misbehaves and does not achieve sealing.*

8.1.1.3 Note-3: De-sealing

Tube-5 had sealed after 134 days (4½ months). Roughly 5½ months after sealing, it de-sealed (almost one year of exposure). After replenishing the saline water, the tube re-sealed in a couple of weeks. One reason for this outlier behavior (de-swelling) may be the loss of water through evaporation from minor leakages, or adsorption of water into the elastomer and gradual water loss from elastomer into air. Another reason, as pointed out in chapter-3, may be the two-way transport of salt and water in an elastomer-brine system, or the breaking and re-forming of polymer crosslink chains. Swelling sometimes decreases a little before building up again. However, from a seal performance viewpoint, it is good to know that *after de-sealing, the elastomer eventually re-seals if exposure to water continues*. *Also, as*

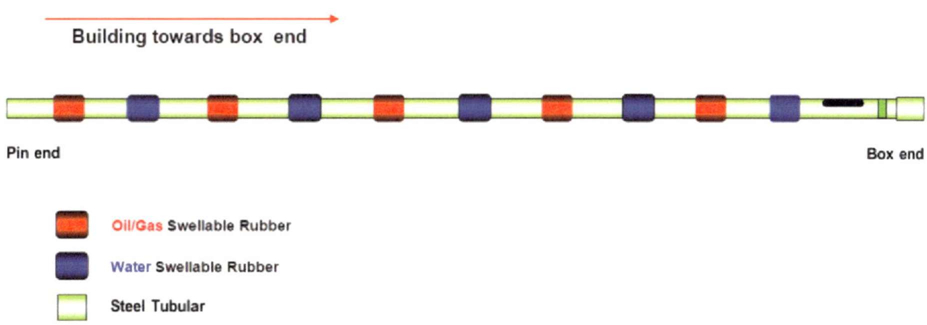

Figure 14.
Hybrid swell packer design; alternate water-swellable and oil-swellable segments on the same steel tubular.

mentioned above, multi-segment approach is a good fail-safe design as it will cover for underperformance of any one elastomer segment.

8.2 Behavior after pressurizing

As described above, four tubes (2, 3, 7, and 8) were gradually pressurized to 1000 psi after they sealed. This was done after a significant post-sealing time in each case, to make sure that the seals were fully intact before pressurizing. Tube-8 was pressurized first, as it was the first one to seal. After observing it under pressure for about 2 weeks, tubes 2, 3, and 7 were also pressurized. By design, pressurization was done on tubes representing all the different test parameters: two packer sizes (3½" and 4½"); three elastomer types (W1, W2, and O1); three swelling media (low and high-salinity water, and oil), and two temperatures (room, and 50°C).

Tube-7 de-sealed within a day of pressurizing to 1000 psi. Pressure was removed from the manifold side. About 2–3 weeks later, the tube re-sealed under atmospheric pressure. After two weeks of resealing, it was pressurized again to 1000 psi. The seal broke again. Later on, after resealing again, it was pressurized to a lower value of 600 psi (40 bar). The tube de-sealed once again. It was not re-pressurized during the remaining test period.

Seals in the other three tubes (2, 3, 8) remained intact after pressurizing to 1000 psi. As mentioned above, a 10–20 bar pressure drop was observed in these tubes within a few days of pressurizing. They were re-pressurized to 1000 psi. The process had to be repeated from time to time.

After about two years of pressurizing, all three tubes de-sealed one after the other. Pressure was initially removed until the tubes re-sealed (about three weeks). When manifold-side pressure was gradually increased, it was found that the sealing withstood a pressure of about 40 bar (600 psi), but not more than that. After about a year (total three years of pressurizing), the seals broke even at this pressure. After another round of de-pressurizing and re-sealing, sealing in tube-2 and tube-3 remained intact at a pressure of about 20 bar (300 psi), but tube-8 de-sealed even at this low pressure. However, it re-sealed once pressure was removed.

8.2.1 Discussion

Performance of tubes 2, 3, and 8 is encouraging for deployment in actual small-to-medium-depth wells, where differential pressure does not exceed 1000 psi. Packers 2 and 3 remained sealed under the full pressure of 1000 psi for two years, then again at 600 psi for one more year, and finally at 300 psi during the last year. Packer-8 performed similarly except for the last year, when its seal broke under pressure. However, it re-sealed when pressure was removed; healthy performance overall.

Behavior of tube-7 was disappointing; it de-sealed every time it was pressurized, even to lower values. Material and condition wise, tubes 3 and 7 were almost similar: W2 rubber, 12% salt solution, 50°C temperature. Tube-7 was a 4½" packer, so the packer-casing gap was smaller, and the seal pressure generated should be higher than tube-3, which was a 3½" packer. All other parameters being the same, if any tube were to de-seal under pressure, it should be tube-3. This anomalous behavior re-strengthens the observation that one individual elastomer segment (of the same material) may underperform due to an outlier issue such as material inconsistency, manufacturing flaw, vulcanizing defect, or some other reason. This also endorses the multi-segment design approach mentioned above; even if one elastomer segment underperforms for any reason, a *serial arrangement of elastomer segments ensures overall sealing*.

9. Chapter summary

This chapter reports the design, fabrication, and commissioning of a test setup for damage assessment and longevity testing (over a 5-year period) of water-swelling and oil-swelling elastomers being used in regional oilfields. Sections of actual packer were placed inside actual steel casings. Sealing performance under low and high-salinity brine and crude oil was investigated at two temperatures, mimicking in-situ conditions in shallow-aquifer type wells and medium-depth wells in the region. Out of ten units in the test facility, nine were actual swell packers inside actual casings, and one transparent Perspex unit was for demonstration purpose. Tube diameter was 3½-in in four 4½-in in five packers. Outer casings were 7-in diameter steel tubulars in all units. Six of the units had water-swelling elastomers, and three had oil-swelling ones. Water-based units were filled with salt-water solutions of low and medium-high salinity (0.5%, and 12%), while the remaining tubes had actual crude oil. The rubber elements in the packers were built up of two different water-swelling elastomers and one oil-swelling elastomer. High pressure (1000 psi) testing was carried in four selected packers, to replicate medium-high well pressures.

Regular log of readings was maintained over the five-year study period. Packers exposed to low salinity and higher temperatures sealed earlier. Water-swelling elastomers sealed faster than oil-swelling ones. Pressurized tubes either retained sealing the whole time, or re-sealed after removing of pressure or reducing it to a lower value. Results obtained are generally in line with swelling elastomer behavior observed in earlier studies. Some failures that have occurred can provide helpful pointers to field engineers and application designers.

Author details

Sayyad Zahid Qamar[1*], Maaz Akhtar[2] and Tasneem Pervez[1]

1 Mechanical and Industrial Engineering Department, Sultan Qaboos University, Muscat, Oman

2 Mechanical Engineering Department, N.E.D. University of Engineering and Technology, Karachi, Pakistan

*Address all correspondence to: sayyad@squ.edu.om

References

[1] Al-Douseri KMM, Barnes C, Young D, Smith PE (2010) "Swellable Packers Provide a Brownfield Water Management Solution in Open and Cased Hole," *SPE Oil and Gas India Conference and Exhibition 2010, OGIC*, Mumbai, India, 20-22 January 2010

[2] Qamar SZ, Pervez T, Akhtar M, Al-Kharusi MSM (2012) "Design and Manufacture of Swell Packers: Influence of Material Behavior," Materials and Manufacturing Processes, 27 (7), 2012, p 727-732, IF 1.629 Oct-2015

[3] Mahrooqi MA, Hinai G, Marketz F (2007) "Improved Well and Reservoir Management in Horizontal Wells Using Swelling Elastomers," SPE Paper # 107882, SPE Annual Technical Conference and Exhibition, 11-14 November 2007, Anaheim, California, USA

[4] Rogers H, Allison D, Webb E (2008) "New Equipment Designs Enable Swellable Technology in Cementless Completions," SPE PAPER # 112302, IADC/SPE Drilling Conference, 4-6 March 2008, Orlando, Florida, USA

[5] Cooper K, McVey A, Schafer D, Cox JD, Hilleary N, Parker D, Crooks J (2008) "Completion of a Horizontal Well with Swellable Packers to Control Water Production," SPE PAPER # 116263-MS, SPE Annual Technical Conference and Exhibition, 21-24 September 2008, Denver, Colorado, USA

[6] Qamar SZ, Hiddabi SA, Pervez T, Marketz F (2009) "Mechanical Testing and Characterization of a Swelling Elastomer," Journal of Elastomers and Plastics, Vol. 41, No. 5, September 2009, p 415-431

[7] Kamal MS, Sultan AS (2019) Enhanced Oil Recovery, in: Jafar MM, Sheardown H, Al-Ahmed A (eds) Functional Polymers, Polymers and Polymeric Composites: A Reference Series, Springer

[8] Zeidan AA, Redha RA, Williams DD, Montero JE (2018) "Implementing WRFM Well, Reservoir, Facility Management and Smart Field Best Practices for EOR Optimization," Paper # IDSPE-193693, SPE International Heavy Oil Conference and Exhibition, 10-12 December 2018, Kuwait City, Kuwait

[9] Hembling D, Salamy S, Qatani A, Carter N, Jacob S (2006) "Swell Packers: Enabling Openhole Intelligent and Multilateral Well Completions for Enhanced Oil Recovery," Proceedings IADC/SPE Asia Pacific Drilling Technology Conference, 13-15 November 2006, Bangkok, Thailand, SPE Paper # 100824-MS

[10] Roland CM (2009) "Vagaries of Elastomer Service Life Predictions," Plastics, Rubber and Composites, 38:8, 349-354

[11] Achenbach M (2000) "Service Life of Seals – Numerical Simulation in Sealing Technology Enhances Prognoses," Computational Materials Science, 19 (1-4), 15 December 2000, p 213-222

[12] Gillen KT, Clough RL (1997) "Prediction of elastomer lifetimes from accelerated thermal-aging experiments," Elastomer-service life prediction symposium, 14-15 Sep 1997Akron, Ohio, USA

[13] Qamar SZ, Pervez T, van de Velden M, Sanchez FJ (2012) "Design and Fabrication of Test Facility for Longevity Testing of Elastomer Seals," ASME-2012 International Mechanical Engineering Congress and Exhibition

(IMECE-2012), Houston, Texas, 9-15 November 2012

[14] Akhtar M, Qamar SZ, Pervez T, Al-Jahwari FK (2018) "Performance Analysis of Swelling Elastomer Seals," Petroleum Science and Engineering, 165 (2018), p 127-135

[15] Qamar SZ, Pervez T, Akhtar M (2016) "Performance Evaluation of Water-Swelling and Oil-Swelling Elastomers," Journal of Elastomers and Plastics, 48 (6) 2016, p 535-545

[16] Qamar SZ, Akhtar M, Pervez T, Al-Kharusi MSM (2013) "Mechanical and Structural Behavior of a Swelling Elastomer under Compressive Loading," Materials and Design, 45 (March 2013), p 487-496

[17] Pervez T, Qamar SZ, van de Velden M (2011) "Comparison between Fresh and Exposed Swelling Elastomer," Journal of Elastomers and Plastics, 44 (3), May 2012, p 237-250

[18] Patel H, Salehi S, Ahmed R, Teodoriu C (2019) "Review of elastomer seal assemblies in oil and gas wells: Performance evaluation, failure mechanisms, and gaps in industry standards," Journal of Petroleum Science and Engineering, 179 (2019), p 1046-1062

Chapter 6

Swelling Elastomers: Comparison of Material Models

Sayyad Zahid Qamar, Maaz Akhtar and Tasneem Pervez

I think all literature should be read as comparative literature.
And I think we should write out of what we know, but in the expectation that
we can be changed at any moment by something we have yet to discover.
Margo Jefferson

Abstract

Little data is available about the material properties and swelling response of the elsatomers used in swell packers. This information is necessary for modeling and simulation of these elastomers in different petroleum applications. An experimental setup was therefore designed and implemented at Sultan Qaboos University (SQU) to investigate the material behavior of these elastomers under tension and compression, so that these properties could be used for different simulations. Before developing a finite element model (FEM) of elastomer seal performance, it was felt that a thorough evaluation needs to be carried out to decide which of the currently available material models is most suitable for swelling elastomers. This comparison translates into the selection of the correct strain energy function for accurate determination of material coefficients. Different hyperelastic material models are compared here. Experimental investigations under tensile and compressive loads, along with their numerical analysis are presented in detail in this chapter.

Keywords: Swelling elastomer, finite element simulation, material model, model comparison

1. Introduction

Metallic and rubber seals have been used for quite some time in oilfield development. However, the advent of swelling elastomer seals has revolutionized the technology. Deployment of swelling elastomers in a variety of petroleum applications are discussed in detail in Chapter 2. Before using swelling elastomers for any application, behavior under different fluids and environmental conditions should be known to predict the actual response. It can be done experimentally, analytically, or through numerical simulations. Though it would be most efficient time-wise, no analytical work is available that entirely explains the behavior of these materials. On the other hand, it is difficult, costly, and time-consuming (and at times even impossible) to perform experiments for all the possible conditions. Hence, a robust numerical simulation strategy, validated against experimental results, can be used to predict the behavior of swelling elastomers.

Little data is available (about their material properties and swelling response), which is needed for modeling and simulation of these elastomers in different petroleum applications. An experimental setup was therefore designed and implemented at Sultan Qaboos University (SQU) to investigate the material behavior of these elastomers under tension and compression, so that these properties could be used for different simulations. Before developing a finite element model (FEM) of elastomer seal performance, it was felt that a thorough evaluation needs to be carried out to decide which of the currently available material models is most suitable for swelling elastomers. This comparison translates into the selection of the correct strain energy function for accurate determination of material coefficients. Different hyperelastic material models are compared here. Experimental investigations under tensile and compressive loads, along with their numerical analysis are presented in detail in this chapter.

2. Models for rubberlike materials

Rubber elasticity theory explains the mechanical properties of a rubber in terms of its molecular constitution. This approach involves two essentially separate issues. First is the treatment of the statistical properties of a single long-chain molecule in terms of its geometrical structure. Second comes the application of this treatment to the problem of the network of long-chain molecules corresponding to a cross-linked or vulcanized rubber.

First statistical mechanics approach to describe the force on a deforming elastomer network was based on Gaussian statistics. The basic assumption in this theory is that a chain never approaches its fully extended length. The problem was initially attempted by Kuhn and Grun [1] who derived a relation between elastic modulus and molecular weight. More precise treatment leading to explicit forms of stress–strain curves valid for large strains were developed by Wall [2], Flory and Rehner [3, 4], and, James and Guth [5]. Results show that these models fail to capture large and even moderate stretches. Hence many constitutive models were developed in an attempt to predict the behavior not only for small stretch levels but also for medium and large stretches.

Other researchers followed non-Gaussian statistics. These are physical models based on an explanation of a molecular chain network, phenomenological invariant-based and stretch-based continuum mechanics approach. The distinctive feature of non-Gaussian approach is that it presumes that a chain can attain its fully extended length. Phenomenological approach is used to relate empirical findings of certain phenomena in such a way that it is consistent with fundamental theory, but is not directly derived from theory. This approach is further divided into stretch-based and invariant-based models. Invariants are basically defined by principal stretches. Models in which stretches can be written in terms of invariants are classified as invariant-based models, while models that cannot be converted into invariants are known as stretch-based models. Micromechanical models describe a material as a three-dimensional network and consider statistical mechanics arguments on networks of cross-linked long-chain molecules. A hyperelastic material model is a type of constitutive relation for rubberlike materials in which the stress–strain relationship is developed from a function. Most continuum mechanics treatment of rubber elasticity begins with assuming rubbers to be hyperelastic and isotropic in nature. **Figure 1** gives a broad classification of different types of hyperelastic material models.

2.1 Micromechanical models

These models assume a unit cell which deforms in a principal-stretch space to relate individual chain stretch to applied deformation. Wang and Guth [6] proposed

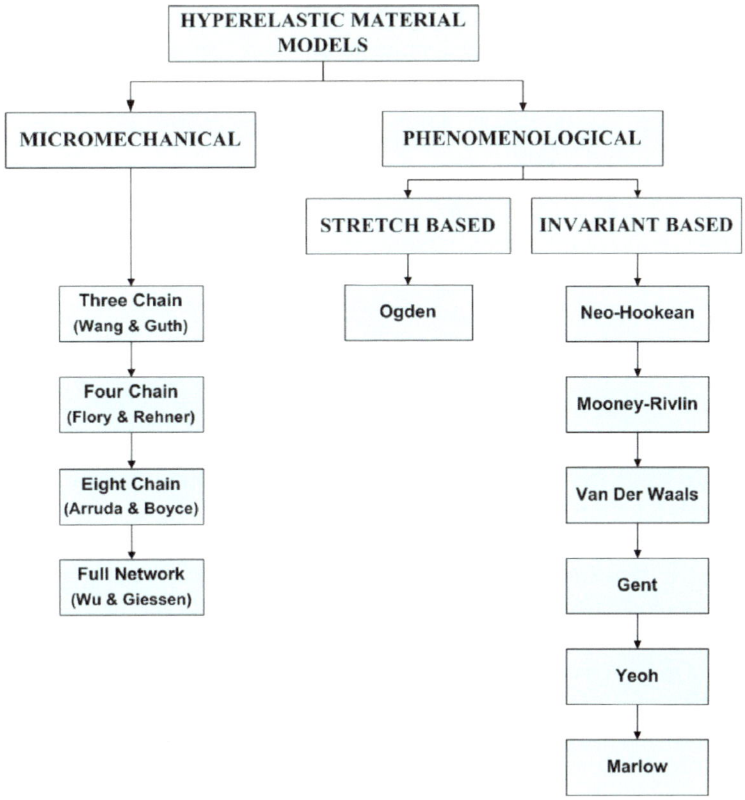

Figure 1.
General classification of different hyperelastic material models.

a three-chain model in which chains are located along the axes of cube. The chains deform affinely with the cell, and stretch of each chain corresponds to the principal stretch value. The strain energy function is given by

$$W_{3-Chain} = \frac{NkT}{3}\sqrt{n}\left[\lambda_1\beta_1 + \sqrt{n}\,ln\left(\frac{\beta_1}{\sinh\beta_1}\right) + \lambda_2\beta_2 + \sqrt{n}\,ln\left(\frac{\beta_2}{\sinh\beta_2}\right) + \lambda_3\beta_3 + \sqrt{n}\,ln\left(\frac{\beta_3}{\sinh\beta_3}\right)\right]$$

$$\beta_i = \mathcal{L}^{-1}\left(\frac{\lambda_i}{\sqrt{n}}\right); i = 1, 2, 3. \tag{1}$$

Here, $\mathcal{L}^{-1}$ is the inverse Langevin, N is the chain density, k is the Boltzman constant, T is the absolute temperature, n is the length of chain, and λ_i are the principal stretches. Three-chain model closely follows experimental data only for uniaxial case, and only for small deformations under shear. Flory and Rehner [3, 4] developed a four-chain tetrahedral model. Later, Treloar [7] attempted to determine the entropy of deformation using numerical method of computation, in which four chains are linked together at the center of a regular tetrahedron. Tetrahedron deforms according to the applied deformation and chains deform in un-affine manner. Four chain (tetrahedral) model gives good match with experimental data for uniaxial and shear cases, but extrapolates beyond the actual stretch region. For biaxial data, three-chain and four-chain models do not yield convincing results. Relationship between the stretches of individual chain to the applied stretch for the four-chain model is determined by an iterative method, hence expression for strain energy function is not provided here.

Arruda and Boyce [8] proposed an eight-chain model in which chains align along diagonals of a unit cell. Due to the symmetry of chain structure, the interior junction point remains at the center throughout the deformation. Strain energy function is given by

$$W_{8-Chain} = \frac{NkT}{3}\sqrt{\mathrm{n}}\left[\lambda_{Chain}\beta_{Chain} + \sqrt{\mathrm{n}}\,ln\left(\frac{\beta_{Chain}}{\sinh\beta_{Chain}}\right)\right]$$

$$\beta_{chain} = \mathcal{L}^{-1}\left(\frac{\lambda_{chain}}{\sqrt{n}}\right)$$

$$\lambda_{chain} = \left[\frac{1}{3}\left(\lambda_1^2 + \lambda_2^2 + \lambda_3^2\right)\right]^{1/2} \tag{2}$$

where, and, λ_{chain} is the chain stretch. Eight-chain model makes good predictions for large strain behavior under different states of deformation, but diverges beyond the actual stretch region for biaxial and shear cases.

Wu and Giessen [9] suggested a full network chain model in which chains are randomly distributed and deform in an affine manner. Strain energy response is determined by overall integration which is computationally intensive; expression for strain energy is therefore reproduced here. To develop constitutive relations, work done by the stresses is established in terms of strain energy function. Strain energy function must depend on the amount of stretch via invariants of the stretch tensor. The coefficients in these functions should be determined by uniaxial, biaxial, and shear test data. The essential problem is to determine the strain energy function for providing good fit with a number of sets of experimental data.

2.2 Phenomenological invariant-based models

One of the first attempts at a hyperelastic material model based on the phenomenological invariant-based approach was made independently by the two scientists Mooney [10] and Rivlin [11]. Due to the assumption of incompressibility, the third invariant is not considered, yielding the following expression for strain energy function:

$$W_{Rivlin} = \sum_{i+j=1}^{\infty} C_{ij}(I_1 - 3)^i (I_2 - 3)^j \tag{3}$$

By keeping only the second term of this expression, we get the famous Mooney-Rivlin model for elastomer deformation

$$W_{Mooney-Rivlin} = C_1(I_1 - 3) + C_2(I_2 - 3) \tag{4}$$

Mooney-Rivlin model makes passable predictions for uniaxial and shear cases in the low-stretch region, but does not yield good predictions for equibiaxial case and uniaxial and shear cases in the large-stretch region. As reported by Treloar [12] and Boyce and Arruda [13], when only the first term of Mooney-Rivlin strain energy function is retained, it becomes the neo-Hookean model:

$$W_{neo-Hookean} = C_{10}(I_1 - 3) \tag{5}$$

Kilian [14] explains that van der Waal hyperelastic material model gives the mathematical relation in terms of the initial shear stress 'μ_s', global interaction parameter '*a*', effective invariant $\check{I} = (1 - \beta_v)I_1 + \beta_v I_2$ and the parameter $\eta_v = \sqrt{\frac{\check{I}-3}{\lambda_m^2-3}}$; β_v is known as the invariant mixture parameter:

$$W = \mu_s \left\{ -(\lambda_m^2 - 3)[ln\,(1 - \eta_v) + \eta_v] - \frac{2}{3} a \left(\frac{\check{I} - 3}{2} \right)^{\frac{3}{2}} \right\} \tag{6}$$

Gent [15] proposed a new constitutive relation for strain energy function which is given by the expression

$$W = -\frac{E}{6} J_m\, ln \left[1 - \left(\frac{J_1}{J_m} \right) \right] \tag{7}$$

where $J_1 = (I_1 - 3)$, and J_m is the maximum value for J_1.

The Yeoh model [16] is derived from Rivlin strain energy function by expanding only the first three terms, and neglecting the second invariant, yielding:

$$W_{Yeoh} = C_{10}(I_1 - 3) + C_{20}(I_1 - 3)^2 + C_{30}(I_1 - 3)^3 \tag{8}$$

The Marlow hyperelastic material model [17] is another variation of this approach:

$$W = C(\lambda_1^2 + \lambda_2^2 + \lambda_3^2) = C(I_1) \tag{9}$$

Here, C is a material parameter, and λ is the principal stretch.

2.3 Phenomenological stretch-based models

Ogden model [18] proposes a strain energy function based on the principal stretches for incompressible materials. Principal stretches are directly measurable quantities and it is one obvious advantage of using them. Ogden strain energy function is given by

$$W = \sum_{n=1}^{Q} \frac{\mu_n}{\alpha_n} \left(\lambda_1^{\alpha_n} + \lambda_2^{\alpha_n} + \lambda_3^{\alpha_n} - 3 \right) \tag{10}$$

where μ_n and α_n are material constants. Nonlinear least square optimization technique is used by Twizell and Ogden [19] to determine the stable constants for the above strain energy function and to find an improve fit to the data when '*n*' increases. This model can be more accurate if data from multiple experiments are available.

There are various other forms of strain energy potentials (**Figure 1**) for modeling hyperelastic isotropic elastomers, such as the ones reported by Ali et al. [20], Hossain and Steinmann [21], and Steinmann et al. [22]. However, these functions seldom describe the complete behavior of these swellable materials, especially for different loading conditions.

2.4 Other models

Some researchers have attempted to develop material models which explain the stress-stretch behavior of an elastomer after swelling to a particular level. As described by Treloar [23], Gauss model follows a molecular approach and involves swelling ratio 'v_s':

$$W = \frac{NkT}{2} v_s^{\frac{1}{3}} \left({\lambda_1}^2 + {\lambda_2}^2 + {\lambda_3}^2 - 3 \right) \tag{11}$$

Increasing discrepancies between Gaussian theory and experiments have been recognized when moving from small to large stretch levels. This discrepancy is amplified when the effect of swelling is included. Flory and Erman [24] model predicts the swelling behavior reasonably well only for small to moderate stretches. Arruda-Boyce model (for stress-stretch behavior coupled with elastomer swelling) is based on a simplified representation of eight chains lying along the diagonal of a cubic cell. The Cauchy stress for Arruda-Boyce model [25] is given by the following relation:

$$\sigma = v_s^{\frac{2}{3}} \frac{NkT}{3} \frac{\sqrt{n}}{\lambda_c} \mathcal{L}^{-1} \left[\left(\frac{v_s^{-\frac{1}{3}} \lambda_c}{\sqrt{n}} \right) \left(\lambda^2 - \frac{1}{\lambda} \right) \right] \tag{12}$$

This model, when used for predicting swelling behavior of elastomers, gives close results only for large stretch. They later attempted to combine their model with Flory-Erman model. This hybrid model is supposed to take into account the actual effect of swelling for small to large stretches. However, even this hybrid model fails to replicate the actual behavior of swelling elastomers.

A few models have been proposed by authors such as Wagner [26]. These are basically hyperelastic models that also include swelling ratio, but fail to give reasonable predictions. There is hence a need for a model that predicts the actual behavior, and includes factors such as energy of mixing, and effect of diffusion, together with hyperelastic effect.

2.5 Energy-diffusion models

Other authors have suggested more realistic constitutive models for elastomers subjected to swelling, in which changes due to configuration entropy are considered along with energy of mixing and diffusion. According to Flory-Rehner theory [3, 4], free energy density can be written as the sum of strain energy density function due to thermodynamics of mixing and configuration entropy of polymer network. A number of authors use this concept to develop a relation for the prediction of swelling, such as Han et al. [27], Cai and Suo [28], Chester and Anand [29], Drozdov and Christiansen [30], Hong et al. [31], and Lucantonio et al. [32].

Flory-Huggins theory, proposed by Flory [33] and Huggins [34], considers the changes in entropy due to mixing. Changes due to configuration entropy are added by considering different available hyperelastic material models as discussed above. As explained by Lucantonio [32], Hong et al. [35], and Kang and Huang [36], solvent influx in swelling elastomer follows the diffusion phenomenon. A few authors, such as Biot [37], Cai et al. [38], Lucantonio and Nardinocchi [39], and Tomari and Doi [40], use the concept of porous media and assume swelling elastomer or gel to be poro-elastic. A porous material is made up of pores and its permeability is related to the pore size. Major portion of a swollen elastomer contains solvent which resists when under the influence of force. That is why models based on poro-elastic concept do not predict good swelling results in general as discussed by Hong et al. [31].

Most of the chemo-mechanical models use Gaussian-chain model for mechanical configuration changes that consider small stretching of chains, and use Fick's law to define fluid influx. Again, they give reasonable results only at small swelling levels. Material models for predicting swelling in polymer networks are discussed below.

2.5.1 Poroelastic models

Biot [37] used Darcy's law along with thermodynamics of mixing to model fluid influx into a porous medium (soil), the approach being known as poro-elasticity theory. This theory is used in a variety of applications to study porous materials from soils to tissues. Tomari and Doi [40] used the concept of poro-elasticity to develop a model for swelling dynamics. They assumed swelling time to be proportional to gel size, frictional constant, and changes in stresses. Barrière and Leibler [41] proposed a swelling model that is based on a porous media concept and includes frictional effect between elastomer and fluid. To describe fluid influx they use the dependence of coefficient of diffusion on concentration, and do not use Fick's law (as according to them it does not predict good behavior at large swelling). Cai et al. [38] performed experiments on swollen gel that is compressed between two plates. They developed a model using linear poro-elasticity for the experiments performed. Variational approach is employed by Baek and Pence [42] to investigate gels in equilibrium, subjected to loading on surface. They do not take into account the relationship for dynamical processes (like diffusion) prior to equilibrium. Lucantonio and Nardinocchi [39] describe bending due to swelling in a gel considering poro-elastic linear theory. Lucantonio et al. [32] investigated the stability between linear poro-elastic and nonlinear theory approaches assuming Gaussian statistics and homogenous configuration. Instead of considering dry state as reference configuration (like other models), they assume initially swollen elastomer as a reference state. Constitutive relations are derived using weak-form variational formulation. Bouklas and Huang [43] present a comparison of linear poro-elasticity and nonlinear Gaussian based (chemo-mechanical) theories. Both theories give similar results at small swelling ratios, but for higher swelling ratios linear theory fails and gives large errors as compared to nonlinear theory.

2.5.2 Chemo-mechanical models

Durning and Morman [44] use phantom network model along with Bastide's scaling model for taking into account the entropy changes due to configuration. Fick's law is used for describing fluid influx. Inconsistencies are present due to the use of Fick's law, as it does not provide a solution for nonlocal effects. Chemo-mechanical theory for hydrogels was developed by Dolbow et al. [45] and extended finite element method was implemented to study swelling. Continuum based sharp interface model and swelling kinetics for temperature sensitive stimulus responsive hydrogels was developed by Ji et al. [46]. Hong et al. [31] assume short and long range migration of solvent inside the polymer. Developed theory is based on non-equilibrium thermodynamics and considers free energy and kinetics to be material specific. They use Flory-Huggins theory to describe changes in free energy, while Gaussian approach is used for defining changes in entropy due to network configuration. They use the diffusion equation for fluid inflow of Feynman et al. [47], in which flux is proportional to gradient of chemical potential. Hong et al. [35] developed an equilibrium theory using Gibb's approach. Legendre transformation is used to represent the field in gel, analogous to compressible hyperelastic solid. Developed model is used on kinetics of fluid influx and Gaussian approach. User-defined subroutine (UHYPER) in ABAQUS is also developed for the model. Duda et al. [48]

follow Gibb's idea of multi-component solid for the development of an expression for swelling equilibrium. Theory is based on mechanics (macro) and chemistry (micro) of mixing of solid and fluid. They consider Gaussian neo-Hookean model for changes in configuration entropy of the elastomer. Development of the model assumes that the body is not fully immersed into the fluid. Kang and Huang [36] formulated a model based on variational approach to set governing equations combining chemical and mechanical conditions. They used similar approach as Hong et al. [31], following Gaussian-chain model for configuration entropy and Flory-Huggins equation for energy of mixing. They also develop an explicit formulation for tangent modulus and true stress. User-defined subroutine (UMAT) in ABAQUS is also developed for numerical investigation and comparison is done with user subroutine developed by Hong et al. [35].

A continuum mechanical theory along with fluid imbibition was developed by Chester and Anand [29]. They first use Gaussian approach to develop a model for swelling phenomenon. Fluid influx is taken into account using Darcy's law. Then they use micro mechanical non-Gaussian based approach for model improvement. As a further extension to their work, they include thermal effect and develop coupled equation for stretching, fluid influx, and heat transfer. Cai and Suo [28] use Flory-Rehner model to describe the changes in entropy due to stretching, but ignore the changes in entropy due to mixing of fluid and polymer. They develop an equation of state based on Gaussian-chain model considering stretching only. Yan and Jin [49] take into account the entanglements of chains, by using hybrid free energy function based on Edwards-Vilgis slip-link model and Flory-Huggins theory. Drozdov and Christiansen [30] propose a swelling model based on viscoelasticity and viscoplasticity. They study hydrogels, under both tensile and compressive loads, to investigate the effect of strain rate on elastic properties. Constitutive relations are based on viscoelasticity of gels.

All of the above models generally give reasonable results only for small swelling. For large swelling (as in the case of swelling elastomers), most of these models do not give close approximation to actual data.

3. Material models for swelling elastomers

None of the existing material models accurately represent the behavior of swelling elastomers. Different types of models for rubberlike materials were briefly described above. Major material models that can be used to represent stretchable elastomers, though not very accurately, are discussed in more detail in this section. As mentioned above, a swelling elastomer can be treated as a hyperelastic material, commonly modeled as an incompressible, homogeneous, isotropic, and nonlinear elastic solid. Due to its long and flexible structure, an elastomer has the ability to stretch to several times its initial length. Elastomers at small strains (up to 10%) have linear stress strain relation, and behave like other elastic materials [15]. In case of applications where large deformations exist, theory of large elastic deformation should be considered. Several theories for large elastic deformation have been developed for hyperelastic materials based on strain energy density functions. These models are based on phenomenological based continuum mechanics approach and micromechanically motivated network approach [20, 21]. Phenomenological models contain invariant-based or principal stretch-based approach usually containing polynomial functions. Micromechanical models typically have terms for cross-linked long-chain molecules [3, 4, 6, 8, 50]. Selection of appropriate strain energy potentials and correct determination of material coefficients are the main factors for modeling and simulation.

3.1 Gaussian Hyperelastic material models

One major approach used to define the deformations in elastomer chains follows Gaussian statistics, Neo-Hookean model being a good example. In Gaussian statistics, a chain never approaches the maximum stretch; rather, it is limited to small-to-moderate stretches. For a three dimensional polymer, due to its large chain density, probability distribution for any event x approaches a Gaussian distribution, given by

$$\mathrm{p}(x) = \frac{1}{\sqrt{2\pi}\sigma_x} \exp\left(-\frac{x^2}{2{\sigma_x}^2}\right). \tag{13}$$

Standard deviation in terms of chain density (N) and distance between the chain ends (r) is given by Treloar [23] as

$$\sigma^2 = \frac{Nr^2}{3}. \tag{14}$$

Probabilities in x, y and z-directions are given by $\mathrm{p}(x)$, $\mathrm{p}(y)$ and $\mathrm{p}(z)$, repectively. Knowing that

$$\mathrm{p(v)} = \mathrm{p}(x)\mathrm{p}(y)\mathrm{p}(z),$$

and assigning $b^2 = \left(\frac{3}{2Nr^2}\right)$, we get the following relation

$$\mathrm{p(v)} = \frac{b^3}{\pi^{3/2}} \exp\left\{-b^2\left(x^2 + y^2 + z^2\right)\right\} = \frac{b^3}{\pi^{3/2}} \exp\left(-b^2r^2\right). \tag{15}$$

Here, (x_0, y_0, z_0) is the initial (unstretched) location, while (x, y, z) are the final coordinates in the stretched condition, and (obviously), $x^2 + y^2 + z^2 = r^2$. According to Boltzmann general principle of thermodynamics, entropy is proportional to the logarithm of the possible configurations corresponding to a specified state. For small volume ($dxdydz$), probability can be used to define entropy (s) as follows:

$$\mathrm{s} = k \ \ln \frac{b^3}{\pi^{3/2}} \exp\left(-b^2r^2\right)dxdydz$$

$$\Rightarrow \mathrm{s} = k\left[\ln \frac{b^3}{\pi^{3/2}} - b^2r^2 + \ln\left(dxdydz\right)\right]$$

$$\Rightarrow, \mathrm{s} = \text{constant} - kb^2r^2 \tag{16}$$

where k is the Boltzmann constant. Taking change in unstretched end-to-end distance of chain $\left({x_o}^2 + {y_o}^2 + {z_o}^2 = {r_o}^2\right)$, and on simplification we get the change in entropy

$$\Delta\mathrm{s} = kb^2\left\{{x_o}^2\left({\lambda_1}^2 - 1\right) + {y_o}^2\left({\lambda_2}^2 - 1\right) + {z_o}^2\left({\lambda_3}^2 - 1\right)\right\}. \tag{17}$$

Taking summation for all chains, substituting $\sum {x_o}^2 = \sum {y_o}^2 = \sum {z_o}^2 = \sum {r_o}^2/3$ into Eq. (17), and on simplification we get

$$\Delta\mathrm{s} = kb^2\left\{{x_o}^2{\lambda_1}^2 - {x_o}^2 + {y_o}^2{\lambda_2}^2 - {y_o}^2 + {z_o}^2{\lambda_3}^2 - {z_o}^2\right\}$$

$$\Rightarrow \Delta s = -kb^2 \frac{r_o^2}{3} \{\lambda_1^2 + \lambda_2^2 + \lambda_3^2 - 3\}$$

$$\Rightarrow . \sum \Delta s = -\frac{kN}{2} (\lambda_1^2 + \lambda_2^2 + \lambda_3^2 - 3) \quad (18)$$

Shear modulus for rubbers and elastomers is given by G=NkT. Helmholtz free energy is given by W $= -T\Delta s$, hence Eq. (18) can be written as

$$W = \frac{G}{2} (I_1 - 3), \quad (19)$$

where I_1 is the first invariant of stretch. Eq. (19) gives the strain energy function for neo-Hookean model [13, 23]. As it is derived using Gaussian statistics, it gives linear response for material where elastomer chain undergoes only small to moderate stretches.

3.2 Non-Gaussian Hyperelastic material models

Hyperelastic material models that follow non-Gaussian statistics consider that chains can stretch to reach maximum extensibility. Many models are formulated based on this approach, such as Mooney-Rivlin [10], Yeoh [16], Arruda-Boyce [8], Ogden [18], Gent [15], etc. Mooney-Rivlin model follows Valanis and Landel theory [51] according to which strain energy function for an isotropic material in finite deformation is of the form

$$W = f(\lambda_1) + f(\lambda_2) + f(\lambda_3). \quad (20)$$

Invariants based on principal stretches are given by

$$I_1 = \lambda_1^2 + \lambda_2^2 + \lambda_3^2,$$
$$I_2 = \lambda_1^2\lambda_2^2 + \lambda_2^2\lambda_3^2 + \lambda_1^2\lambda_3^2,$$

and

$$I_3 = \lambda_1^2\lambda_2^2\lambda_3^2. \quad (21)$$

For an incompressible isotropic material, third invariant vanishes and strain energy becomes a function of I_1 and I_2 only:

$$W(\lambda_1, \lambda_2, \lambda_3) = C_1(\lambda_1^2 + \lambda_2^2 + \lambda_3^2 - 3) + C_2\left(\frac{1}{\lambda_1^2} + \frac{1}{\lambda_1^2} + \frac{1}{\lambda_1^2} - 3\right)$$
$$W = C_1(I_1 - 3) + C_2(I_2 - 3). \quad (22)$$

Eq. (22) gives the strain energy function for Mooney-Rivlin model, where C_1 and C_2 are material constants. According to this theory, stress–strain behavior depends upon the partial derivatives $\left(\frac{\partial W}{\partial I_1}, \frac{\partial W}{\partial I_2}\right)$. Characterization of elastic properties of a polymer consists of determining these partial differentials through experimental measurements. These partial differentials cannot be determined from experiments with only one deformation mode. Preferred methods are generally biaxial extension where stretches in both directions vary independently. Yeoh assumed that $\partial W/\partial I_1$ is much larger than $\partial W/\partial I_2$, so the second partial derivative can be neglected. Strain energy function for Rivlin [11] can then be written as

$$\mathrm{W} = \sum_{i+j=1}^{\infty} \mathrm{C}_{ij}(I_1 - 3)^i (I_2 - 3)^j$$
$$\Rightarrow, \mathrm{W} = \sum_{i=1}^{\infty} \mathrm{C}_{ij}(I_1 - 3)^i \tag{23}$$

where index 'j' will always be zero. If the series is truncated after three terms, we get the expression for strain energy function for Yeoh model [16]:

$$\mathrm{W} = \mathrm{C}_{10}(I_1 - 3) + \mathrm{C}_{20}(I_1 - 3)^2 + \mathrm{C}_{30}(I_1 - 3)^3 \tag{24}$$

where C_{10}, C_{20} and C_{30} are material coefficients determined by fitting experimental data. If Eq. (23) is expanded only for '$i = 1$', we get the Neo-Hookean strain energy function, in which $\mathrm{C}_{10} = G/2$.

Another stretch based non-Gaussian phenomenological strain energy function for hyperelastic materials was proposed by Ogden [18]. This model is given by the following relation in terms of principal stretches:

$$\mathrm{W} = \sum_{n=1}^{Q} \frac{\mu_n}{\alpha_n}\left(\lambda_1^{\alpha_n} + \lambda_2^{\alpha_n} + \lambda_3^{\alpha_n} - 3\right). \tag{25}$$

Here, 'μ_n' and 'α_n' are material constants that are determined by fitting the experimental data, and 'Q' is a positive definite integer. Material constants are related to shear modulus (G) by the following relation:

$$\sum_{n=1}^{Q} \mu_n \alpha_n = 2\mathrm{G}. \tag{26}$$

Second degree Ogden strain energy function is given by expanding Eq. (24) for '$Q = 2$':

$$\mathrm{W} = \frac{\mu_1}{\alpha_1}\left(\lambda_1^{\alpha_1} + \lambda_2^{\alpha_1} + \lambda_3^{\alpha_1} - 3\right) + \frac{\mu_2}{\alpha_2}\left(\lambda_1^{\alpha_2} + \lambda_2^{\alpha_2} + \lambda_3^{\alpha_2} - 3\right),$$
$$\mu_1\alpha_1 + \mu_2\alpha_2 = 2\mathrm{G}. \tag{27}$$

It is interesting to note that if Ogden model Eq. (24) is expanded for '$\mathrm{Q} = 2$' and for $\alpha_1 = 2$, $\alpha_2 = -2$, it gives an expression similar to the Mooney-Rivlin model:

$$\mathrm{W} = \frac{\mu_1}{2}\left(\lambda_1^2 + \lambda_2^2 + \lambda_3^2 - 3\right) - \frac{\mu_2}{2}\left(\lambda_1^{-2} + \lambda_2^{-2} + \lambda_3^{-2} - 3\right)$$
$$\Rightarrow .\mathrm{W} = \frac{\mu_1}{2}(I_1 - 3) - \frac{\mu_2}{2}(I_2 - 3) \tag{28}$$

Also, if Ogden model is expanded for '$\mathrm{Q} = 1$' with $\alpha_1 = 2$, $\alpha_2 = 0$, it reduces to the neo-Hooken model:

$$\mathrm{W} = \frac{\mu_1}{2}\left(\lambda_1^2 + \lambda_2^2 + \lambda_3^2 - 3\right) = \frac{\mu_1}{2}(I_1 - 3). \tag{29}$$

Arruda-Boyce model [8] uses the statistical mechanics approach followed by many researchers [3, 4, 6, 50]. Instead of the 3-chain [6] and 4-chain models [3, 4] of earlier researchers, Arruda-Boyce proposed a more accurate 8-chain approach, yielding the following strain energy function:

$$\mathrm{W} = NkT\sqrt{n}\left[\beta_{chain}\lambda_{chain} + \sqrt{n}\ln\left(\frac{\beta_{chain}}{sinh\,\beta_{chain}}\right)\right]. \tag{30}$$

Here, $\beta_{chain} = \mathcal{L}^{-1}\left(\frac{\lambda_{chain}}{\sqrt{n}}\right)$ is the inverse Langevin, N is the chain density, k is the Boltzman constant, T is the temperature, n is the chain length, and λ_{chain} is the chain stretch. Also, $\mathrm{G} = NkT$ is the effective shear modulus.

Since $\mathcal{L}(x) = coth(x) - \frac{1}{x}$, and $coth(x) = \frac{1}{x} + \frac{1}{3}x - \frac{1}{45}x^3 + \frac{2}{945}x^5\cdots$, we can get

$$\mathcal{L}(x) = \frac{1}{3}x - \frac{1}{45}x^3 + \frac{2}{945}x^5\cdots$$

Also,

$$\mathcal{L}^{-1}(x) = 3x - \frac{9}{5}x^3 + \frac{297}{175}x^5\cdots$$

and

$$sinh\,(x) = x + \frac{1}{6}x^3 + \frac{1}{20}x^5 - \frac{1}{5040}x^7\cdots \tag{31}$$

By showing that the expression for chain stretch reduces to a function of the first stretch invariant, $\lambda_{chain} = \sqrt{I_1/3}$, and using Eqs. 4.30 and 4.31, the final expression for Arruda-Boyce strain energy function becomes

$$\mathrm{W} = \mathrm{G}\left[\frac{1}{2}(\mathrm{I_1} - 3) + \frac{1}{20\mathrm{N}}(\mathrm{I_1}^2 - 9) + \frac{11}{1050\mathrm{N}^2}(\mathrm{I_1}^3 - 27) + \frac{19}{7000\mathrm{N}^3}(\mathrm{I_1}^4 - 81) + \cdots\right]. \tag{32}$$

Gent [15] proposed an empirical constitutive relation using only two constants for the entire range of strains for hyperelastic materials. In statistical mechanics approach, when a chain reaches the fully stretched state, it is known as locking stretch. Gent describes this stretch as a maximum value of $\mathrm{J_1}$ denoted by J_m, so that the strain energy function is given by the following relation:

$$\mathrm{W} = -\frac{E}{6}\mathrm{J}_m\,ln\left[1 - \left(\frac{\mathrm{J_1}}{\mathrm{J}_m}\right)\right]. \tag{33}$$

Boyce [52] performed a comparison between Gent and Arruda-Boyce eight chain models. As $\mathrm{J_1} = (I_1 - 3)$ and $I_1 = \lambda_1^2 + \lambda_2^2 + \lambda_3^2$, Gent model can be expanded using natural logarithm.

$$\mathrm{W} = \frac{E}{6}\mathrm{J}_m\left[\frac{\mathrm{J_1}}{\mathrm{J}_m} + \frac{1}{2}\left(\frac{\mathrm{J_1}}{\mathrm{J}_m}\right)^2 + \frac{1}{3}\left(\frac{\mathrm{J_1}}{\mathrm{J}_m}\right)^3 + \cdots + \frac{1}{\mathrm{n}+1}\left(\frac{\mathrm{J_1}}{\mathrm{J}_m}\right)^{\mathrm{n}+1}\right]$$

$$\Rightarrow \mathrm{W} = \frac{E}{6}\left[(I_1 - 3) + \frac{1}{2\mathrm{J}_m}(I_1 - 3)^2 + \frac{1}{3{\mathrm{J}_m}^2}(I_1 - 3)^3 + \cdots\right]$$

$$\Rightarrow .\mathrm{W} = \sum_{i=1}^{\infty}\mathrm{C}_i(I_1 - 3)^i \tag{34}$$

Similarly, Arruda-Boyce model can be re-written in terms of general invariant based form as follows

$$W = \sum_{i=1}^{\infty} C_i \left(I_1^i - 3^i \right). \tag{35}$$

Eqs. 4.34 and 4.35 can be used to predict large stretch deformations.

There are many hyperelastic material models available. Some of them are presented in this section, while others were discussed in the previous section. Models such as Van der Waal and Marlow models have not been presented here because they do not show good convergence for swelling elastomers. Next section gives the comparison of different hyperelastic material models against experimental results from uniaxial tensile tests.

4. Comparison of material models under tensile loading

As none of the available material models accurately represents swelling elastomers, modeling of these elastomers based on current models can yield only approximate results. Before this, no evaluation of material models for prediction of swelling elastomer behavior could be found in published literature. This study was therefore conducted to have a comparative assessment of current hyperelastic material models; how closely they predict the actual behavior under tension. Two water swelling elastomers are used for experimental and numerical investigation under tension (material-*TA* and material-*TB*). Tests are carried out in accordance with ASTM-D412 standard test method [53]. Tensile tests are also simulated, using the commercial finite element package ABAQUS [54]. Simulations are done using different hyperelastic material models available in the ABAQUS materials library. Material coefficients for each model are determined from experimental results using the curve fitting procedure of ABAQUS.

4.1 Experimental work

The main objective of the experimental work is to provide actual tensile test data to extract parameters for the material models, and then to compare predicted behavior from each model against the actual tensile behavior. A large number of samples of two different water-swelling elastomers were allowed to swell for 8 days in saline water of 40,000 ppm (4%) concentration at a temperature of 60°C (actual field conditions in a regional oil-well). Tensile tests were carried out on ring samples before and after swelling, reported values being average of readings from three samples. As expected, the material expands due to swelling. Dimensions of the ring samples before and after swelling are shown in **Figure 2**.

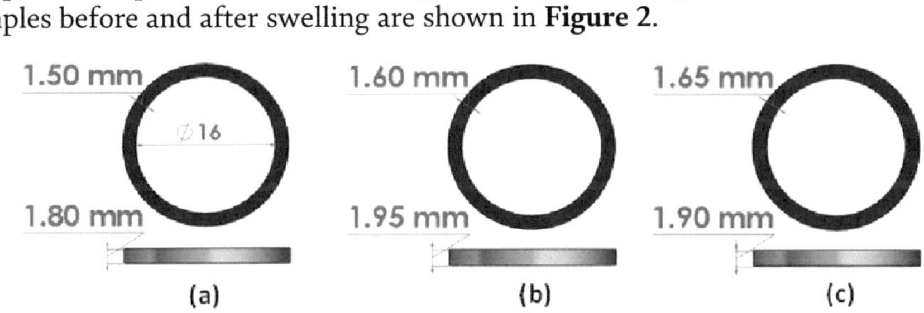

Figure 2.
Samples before and after swelling: (a) before swelling, (b) material TA *after swelling, (c) material* TB *after swelling.*

4.1.1 Sample preparation

Ring samples were cut in line with ASTM-D412, using a die-and-punch set; outside diameter (d_o) 19 *mm*, inside diameter (d_i) 16 *mm*, radial width (w) 1.5 *mm*, and thickness (t) 1.8 *mm*. Actual identification numbers of the two swelling-elastomer materials are not mentioned here due to reasons of confidentiality; they are simply referred to as material *TA* and material *TB*.

4.1.2 Test procedure

Width, thickness, and flattened length of all ring samples (before and after swelling) are recorded prior to conducting the tensile test, to be used later for stress and strain measurements. Specially fabricated (in-house) hook-type fixtures (in accordance with ASTM D412) are used to grip the ring samples. A small pre-tension is applied on each sample to get the flattened length as shown in **Figure 3**. A loading rate of 500 mm/min is used until the sample ruptures. Force-elongation and stress–strain readings are automatically recorded by the machine.

4.2 Numerical Modeling and simulation

Tensile tests are modeled and simulated using the commercial finite element package ABAQUS. Standard dumbbell shaped tensile specimen is used for simulation because of its symmetric nature, allowing modeling of only one-quarter of the sample; **Figure 4**. Load application is also simplified because of this sample geometry, yielding proper uni-axial tension. Swelling elastomer specimen is modeled as a hyperelastic body using 8-noded linear brick element with reduced integration (C3D8R). Deformed specimen is also shown in **Figure 4**. Minor inconsistencies between experimental and simulated results can be expected, as the sample geometries (ring-type and dumbbell-type) are different.

Four well-known material models (Ogden, Yeoh, Arruda-Boyce, and neo-Hookean) are selected from the pool of hyperelastic models available in ABAQUS. These are known to show convergence and good agreement with experimental data for elastomeric materials under tension. Coefficients for each model are extracted from the uniaxial tensile test data (experimental) using curve fitting techniques. Poisson's ratio for these elastomers has been taken as $\upsilon = 0.495$, while density values (from actual experiments) for un-swelled (US) and swelled (S) specimens of materials *TA* and *TB* are $\rho_{US-TA} = 1170\ \text{kg/m}^3$, $\rho_{US-TB} = 1187\ \text{kg/m}^3$, $\rho_{S-TA} = 1282\ \text{kg/m}^3$, and $\rho_{S-TB} = 1293\ \text{kg/m}^3$, respectively.

Figure 3.
Tensile test setup using elastomer ring samples; different stages of testing.

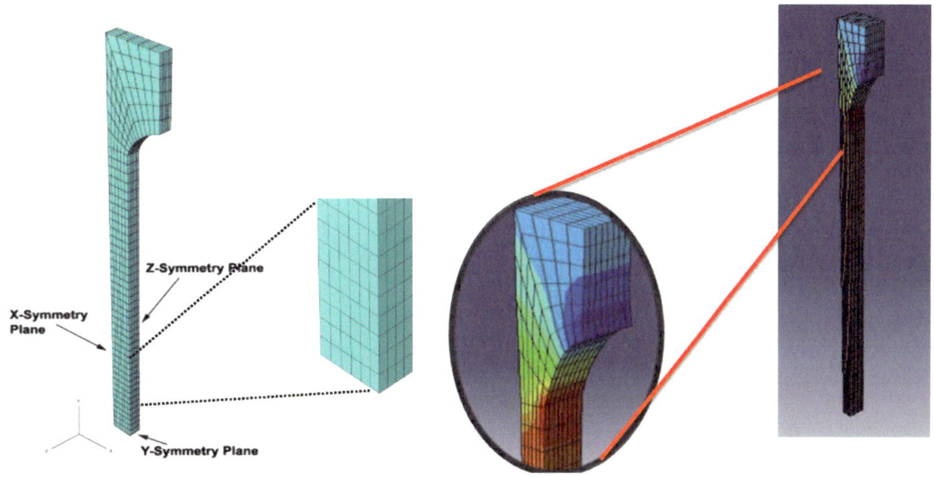

Figure 4.
Finite element model, and deformed shape, of dumbbell type test specimen.

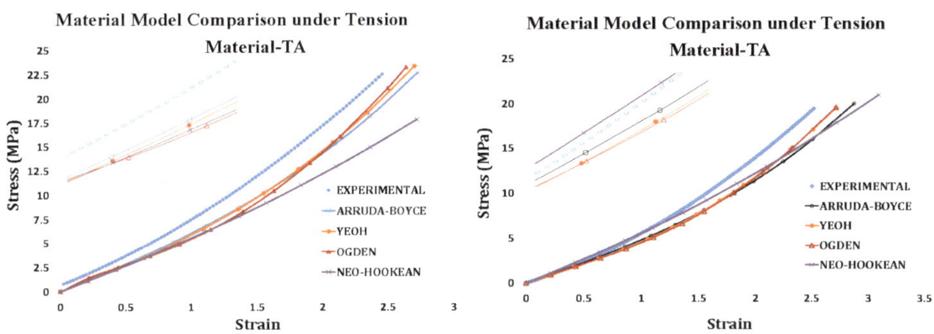

Figure 5.
Stress–strain curves for experimental and FE simulation results for material TA*; before swelling (left); after 8 days of swelling (right).*

4.3 Results and discussion

For both materials, finite element simulation of uni-axial tensile test is done using the four most popular material models for the un-swelled and 8-day swelled conditions. Stress–strain curves from simulations, using the selected material models, are compared with experimental results. Results are plotted separately for materials *TA* and *TB*. During the tensile tests, elastomer *TB* exhibited more stiffness, fracturing at much lower strains than elastomer *TA*.

Figure 5 shows comparison of stress–strain curves for experimental and FE simulation data for un-swelled and swelled (8 days) conditions for material *TA*. In the unswelled condition, at small deformations, all models predict almost the same behavior, Ogden model giving minimum error. For small to medium deformation, Arruda-Boyce model gives minimum error. For large deformations, Ogden model gives minimum error. After 8 days of swelling, for small deformations, all the models again predict almost the same behavior, Arruda-Boyce and neo-Hookean models yielding minimum error. For small to medium deformations, Neo-Hookean model gives the closest results. For large deformations, Ogden model again yields minimum error, Neo-Hookean model giving much larger error.

For material *TB*, experimental and simulated stress–strain curves for un-swelled and 8-day-swelled samples are shown in **Figure 6**. Before swelling, and at small

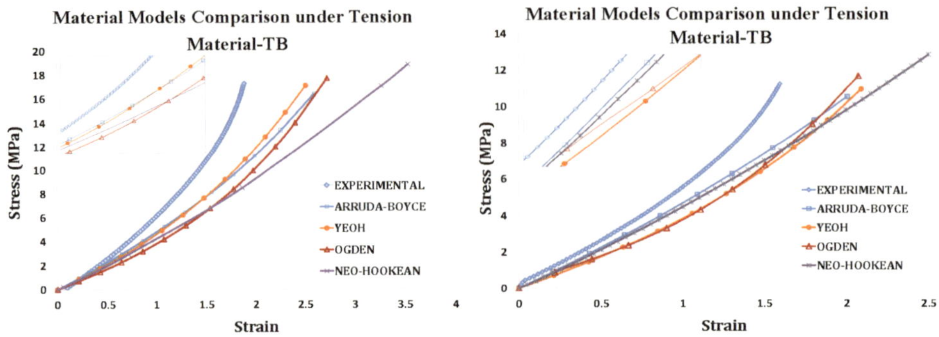

Figure 6.
Stress–strain curves for experimental and FE simulation results for material TB*; before swelling (left); after 8 days of swelling (right).*

deformations, all models are very close to each other, Arruda-Boyce giving the best fit. The trend remains the same for small to medium deformations. For large deformations, Yeoh model gives the closest results. After swelling for 8 days, Arruda-Boyce model shows the closest fit to actual data for small deformations, the other models giving slightly higher error. The trend continues for small to medium deformations. For large deformations, Arruda-Boyce curve starts to diverge away from experimental results while Ogden and Yeoh models show converging tendency, Ogden model yielding the minimum error.

Based on results of the current study, the neo-Hookean model appears to be the best suited for both un-swelled and swelled (small to medium stretch) conditions under tensile loads, but it starts diverging away from experimental results for very large strains. For large strains (which is the case in real-world use of swelling elastomer seals), Ogden model is consistently better than the others. However, it can be seen that no material model predicts results close to the experimental values. The reason might be that energy and diffusion effects are not considered in these models.

5. Comparison of material models under compressive loading

Petroleum industry uses swelling elastomers in a variety of applications as discussed in detail in Chapter 2. Being primarily used as a sealing element, swelling elastomers generally experience compressive loads. Hence, investigation of such elastomers under compression is very important. A comparison of currently available hyperelastic material models is presented in this section against experimental results for two swelling elastomers (*A* and *B*) under compressive loading, before and after different stages of swelling. Just as for tensile loading, such an evaluation of material models for swelling elastomers under compression was not available in published literature before this study. Tests are carried out using disc samples in accordance with ASTM-D575 standard test method [55]. Compression tests are also simulated using the different material models available in the commercial finite element package ABAQUS [54]. Material coefficients for each model are determined from experimental stress–strain data using the curve fitting procedure of ABAQUS.

5.1 Experimental work

To replicate actual field conditions in many regional oil wells, Qamar et al. [56, 57] used samples of two types of water-swelling elastomers in various petroleum development applications were subjected to swelling in saline water of 0.6% (6,000 ppm) and 12% (120,000 ppm) concentrations maintained at a temperature of 50°C. To investigate the effect of swelling, samples were taken out for mechanical testing after 1, 2, 4, 7, 10, 16, 23, and 30 days of swelling. The elastomers were of a fast-swell type, so readings were initially taken after almost each day, and on a weekly basis later on.

5.1.1 Sample preparation

Disc samples were cut using a die-and-punch set, with some surface grinding needed for final trimming. As prescribed by ASTM-D575, disc dimensions for compression testing are 28.5 mm diameter and 12.5 mm thickness. Due to reasons of confidentiality, actual identification number of swelling-elastomer materials is not mentioned here.

5.1.2 Test procedure

Compression tests are performed on a Tinius Olsen universal testing machine in the compression mode, using a 50-kN load cell; **Figure 7**. Disc specimen is placed on a fixed bottom plate while the top plate applies a compressive load on the specimen. Elastomer sample is free to expand in the radial direction. Top surface of the disc moves downward with the compression load, while bottom surface is not allowed to move in the axial direction. Load is applied at a rate of 12 mm/min (ASTM-D575) until the specimen thickness is compressed to 10 mm. Force-deformation and stress–strain data are recorded. A small barreling effect can be observed in the compressed sample.

5.2 Numerical Modeling and simulation

Finite element analysis for predicting the deformation behavior of swelling elastomers under compression for both the materials in two salinities solutions are performed using commercial finite element analysis package ABAQUS [54]. Swelling elastomer specimen is modeled using 8-noded linear brick element with reduced integration (C3D8R). Stress–strain data obtained from experiments are used to extract coefficients that define the constitutive relation in finite element analysis. Material parameters such as Poisson's ratio and density at each condition with swelling for different days are also determined experimentally. ASTM standard disc

Figure 7.
Setup for compression testing, using disc samples.

geometry is modeled for simulation of compression tests. Compressive loads and boundary conditions are in line with experimental values. Simulations are conducted before swelling and after each swelling period for a total of one month period under both salinities. **Figure 8** shows the deformed and undeformed specimens.

5.3 Results and discussion

Finite element simulations of compression experiments are carried out for comparison of the five most suitable hyperelastic material models, including Ogden model with both first and second strain energy functions. Stress–strain curves obtained through simulations (for both the materials) are compared against experimental results after each swelling period.

Experimental and numerical stress–strain curves are compared in **Figure 9**, for material *A* in high salinity solution after 23 days of swelling. Apart from the five models shown in the figure, other material models failed to converge. For small deformations (10–15%), all models give quite good prediction. However, for medium to large deformations, none of the models is even remotely close to experimental results. Compared to the others, Yeoh and Ogden model with second strain energy function (N = 2) are relatively closer to the experimental curve, Ogden being the better one.

Experimental stress–strain curves for material *B* are compared with simulation results in **Figure 10** for low salinity solution after 30 days of swelling. Arruda-Boyce model failed to converge. Neo-Hookean and Ogden (with first strain energy potential) models are nowhere close to the experimental results. Yeoh and Ogden (strain energy potential of second degree) models give somewhat better results for small to moderate deformations. None of the models is good at large strains.

Experimental investigation of performance of swelling elastomer seals (such as sealing pressure at various stages of swelling) is highly difficult and very costly. Numerical simulations, if yielding reasonable predictions, can be a much better alternative. If we want to conduct numerical studies of elastomer seal performance, Ogden model with second-degree strain energy function will have to be used as it gives overall best results both under tensile and compressive loads. This best available current hyperelastic model (Ogden-2) has been later used to study the mechanical and structural behavior of swelling elastomers under compressive

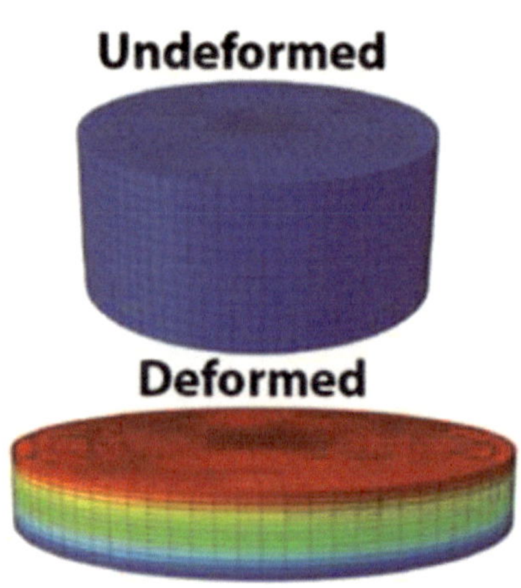

Figure 8.
Specimen (FE simulation) before and after deformation.

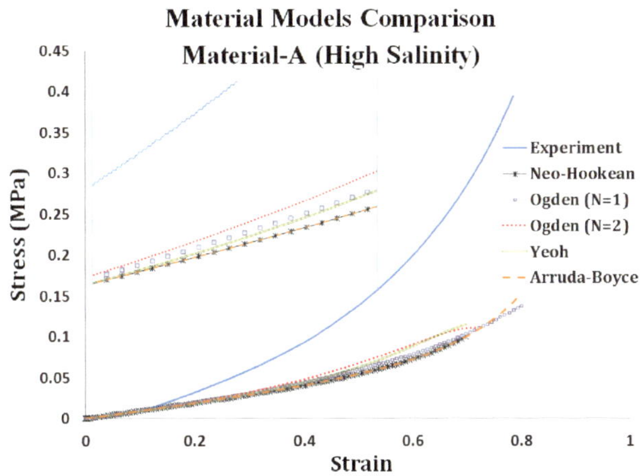

Figure 9.
Comparison of experimental and simulated (different hyperelastic material models) stress–strain behavior under compression for material A *in high salinity after 23 days of swelling.*

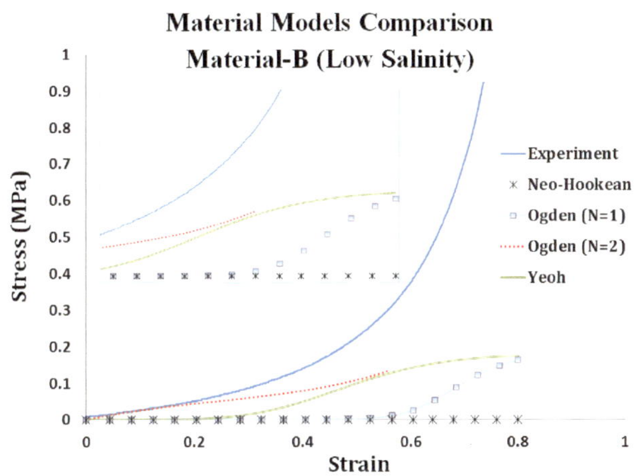

Figure 10.
Comparison of experimental and simulated (different hyperelastic material models) stress–strain behavior under compression for material B *in low salinity after 30 days of swelling.*

loading in Chapter 7, and to evaluate the performance of swelling elastomer seals in Chapter 8.

To more accurately predict the behavior of swelling elastomers, thermodynamics of mixing and the phenomenon of diffusion should also be considered. Starting with the Ogden model, this approach for mechanics of swelling is followed in Chapter-6 to develop a new material model for swelling elastomers.

6. Conclusions

Simulations have been carried out to compare different hyperelastic material models against experimental results for swelling elastomers subjected to tensile and compressive loading, including Arruda-Boyce, Ogden (N = 1), Ogden (N = 2), Yeoh, and neo-Hookean models. Following ASTM standards, experiments were conducted on ring samples (for tensile tests) and disc samples (for compression tests), before swelling, and after 8 days of swelling, using two swelling elastomer materials. Simulations were performed using the FEA package ABAQUS. For tensile loads, most of the models give reasonable predictions for small deformations. For medium to large deformations, Ogden model with second strain energy function (Ogden-2) gives best results. For simulations under compression, many models do not give good predictions even for small loads. Ogden-2 again yields overall minimum error. Hyperelastic models can be used for FE analysis of swellables, giving reasonably acceptable approximations. However, a new material model is definitely needed to capture the material behavior of swelling elastomers more accurately.

Author details

Sayyad Zahid Qamar[1]*, Maaz Akhtar[2] and Tasneem Pervez[1]

1 Mechanical and Industrial Engineering Department, Sultan Qaboos University, Muscat, Oman

2 Mechanical Engineering Department, N.E.D. University of Engineering and Technology, Karachi, Pakistan

*Address all correspondence to: sayyad@squ.edu.om

References

[1] Kuhn W, Grün F (1942) "Beziehungen Zwischen Elastischen Konstanten und Dehnungsdoppelbrechung Hochelastischer Stoffe," Kolloid-Zeitschrift, 101(3), 248-271.

[2] Wall FT (1942) "Statistical Thermodynamics of Rubber," Rubber chemistry and technology, 15(3), 468-472.

[3] Flory PJ, Rehner Jr J (1943) "Statistical Mechanics of Cross-Linked Polymer Networks: Rubberlike Elasticity," Journal of Chemical Physics, 11, 512-520.

[4] Flory P, Rehner J (1943) "Statistical Mechanics of Cross-Linked Polymer Networks II. Swelling," Journal of Chemical Physics, 11, 521-526.

[5] James H, Guth E (1943) "Theory of the Elastic Properties of Rubber," Journal of Chemical Physics, 11(10), 455-481.

[6] Wang MC, Guth E (1952) "Statistical Theory of Networks of Non-Gaussian Flexible Chains," Journal of Chemical Physics, 20, 1144-1157.

[7] Treloar L (1943) "The Elasticity of a Network of Long-Chain Molecules — II," Transactions of the Faraday Society, 39, 241-246.

[8] Arruda EM, Boyce MC (1993) "A Three-Dimensional Constitutive Model for the Large Stretch Behavior of Rubber Elastic Materials," Journal of the Mechanics and Physics of Solids, 41(2), 389-412.

[9] Wu P, Van der Giessen E (1992) "On Improved 3-D Non-Gaussian Network Models for Rubber Elasticity," Mechanics Research Communications, 19(5), 427-433.

[10] Mooney M (1940) "A Theory of Large Elastic Deformation," Journal of Applied Physics, 11, 582-592.

[11] Rivlin, R. (1948). Large elastic deformations of isotropic materials. Iv. Further developments of the general theory. Philosophical Transactions of the Royal Society of London. Series A, Mathematical and Physical Sciences, 241 (835), 379-397.

[12] Treloar L (1944) "Stress-Strain Data for Vulcanised Rubber under Various Types of Deformation," Transactions of the Faraday Society, 40, 59-70.

[13] Boyce MC, Arruda EM (2000) "Constitutive Models of Rubber Elasticity: A Review," Rubber chemistry and technology, 73(3), 504-523.

[14] Kilian H-G (1980) "A Molecular Interpretation of the Parameters of the Van der Waals Equation of State for Real Networks," Polymer Bulletin, 3(3), 151-158.

[15] Gent A (1996) "A New Constitutive Relation for Rubber," Rubber Chemistry and Technology, 69(1), 59-61.

[16] Yeoh O, Fleming P (1997) "A New Attempt to Reconcile the Statistical and Phenomenological Theories of Rubber Elasticity," Journal of Polymer Science-B Polymer Physics Edition, 35(12), 1919-1932.

[17] Marlow R (2003) "A General First-Invariant Hyperelastic Constitutive Model," Constitutive Models for Rubber, 157-160.

[18] Ogden R (1972) "Large Deformation Isotropic Elasticity — On the Correlation of Theory and Experiment for Incompressible Rubberlike Solids," Proceedings of the Royal Society of London. A.

Mathematical and Physical Sciences, 326 (1567), 565-584.

[19] Twizell E, Ogden R (1983) "Non-Linear Optimization of the Material Constants in Ogden's Stress-Deformation Function for Incompressinle Isotropic Elastic Materials," The Journal of the Australian Mathematical Society. Series B. Applied Mathematics, 24(04), 424-434.

[20] Ali A, Hosseini M, Sahari B (2010) "A Review of Constitutive Models for Rubber-like Materials," American Journal of Engineering and Applied Sciences, 3(1), 232.

[21] Hossain M, Steinmann P (2013) "More Hyperelastic Models for Rubber-Like Materials: Consistent Tangent Operators Aad Comparative Study," Journal of the Mechanical Behavior of Materials, 22(1-2), 27-50.

[22] Steinmann P, Hossain M, Possart G (2012) "Hyperelastic Models for Rubber-Like Materials: Consistent Tangent Operators and Suitability for Treloar's Data," Archive of Applied Mechanics, 82(9), 1183-1217.

[23] Treloar LRG (1975) The physics of Rubber Elasticity, Oxford University Press.

[24] Flory PJ, Erman B (1982) "Theory of Elasticity of Polymer Networks," Macromolecules, 15(3), 800-806.

[25] Boyce MC, Arruda EM (2001) "Swelling and Mechanical Stretching of Elastomeric Materials," Mathematics and Mechanics of Solids, 6(6), 641-659.

[26] Wagner M (1994) "The Origin of the C2 Term in Rubber Elasticity," Journal of Rheology (1978-present), 38 (3), 655-679.

[27] Han WH, Horka F, Mckenna GB (1999) "Mechanical and Swelling Behaviors of Rubber: A Comparison of Some Molecular Models with Experiment," Mathematics and Mechanics of Solids, 4(2), 139-167.

[28] Cai S, Suo Z (2012) "Equations of State for Ideal Elastomeric Gels," EPL (Europhysics Letters), 97(3), 34009.

[29] Chester SA, Anand L (2010) "A Coupled Theory of Fluid Permeation and Large Deformations for Elastomeric Materials," Journal of the Mechanics and Physics of Solids, 58(11), 1879-1906.

[30] Drozdov A, Christiansen J (2013) "Constitutive Equations in Finite Elasticity of Swollen Elastomers," International Journal of Solids and Structures, 50(9), 1494-1504.

[31] Hong W, Zhao X, Zhou J, Suo Z (2008) "A Theory Of Coupled Diffusion And Large Deformation In Polymeric Gels," Journal of the Mechanics and Physics of Solids, 56(5), 1779-1793.

[32] Lucantonio A, Nardinocchi P, Teresi L (2013) "Transient Analysis of Swelling-Induced Large Deformations in Polymer Gels," Journal of the Mechanics and Physics of Solids, 61(1), 205-218.

[33] Flory PJ (1942) "Thermodynamics of High Polymer Solutions," Journal of Chemical Physics, 10, 51-61.

[34] Huggins ML (1942) "Some Properties of Solutions of Long-Chain Compounds," The Journal of Physical Chemistry, 46(1), 151-158.

[35] Hong W, Liu Z, Suo Z (2009) "Inhomogeneous Swelling of a Gel in Equilibrium with a Solvent and Mechanical Load," International Journal of Solids and Structures, 46(17), 3282-3289.

[36] Kang MK, Huang R (2010) "A Variational Approach and Finite Element Implementation for Swelling of Polymeric Hydrogels under Geometric

Constraints," Journal of Applied Mechanics, 77(6), 061004.

[37] Biot MA (1941) "General Theory of Three-Dimensional Consolidation," Journal of Applied Physics, 12, 155-164.

[38] Cai S, Hu Y, Zhao X, Suo Z (2010) "Poroelasticity of a Covalently Crosslinked Alginate Hydrogel under Compression," Journal of Applied Physics, 108(11), 113514.

[39] Lucantonio A, Nardinocchi P (2012) "Reduced Models of Swelling-Induced Bending of Gel Bars," International Journal of Solids and Structures, 49(11), 1399-1405.

[40] Tomari T, Doi M (1994) "Swelling Dynamics of a Gel undergoing Volume Transition," Journal of the Physical Society of Japan, 63(6), 2093-2101.

[41] Barrière B, Leibler L (2003) "Kinetics of Solvent Absorption and Permeation through a Highly Swellable Elastomeric Network," Journal of Polymer Science Part B: Polymer Physics, 41(2), 166-182.

[42] Baek S, Pence T (2011) "Inhomogeneous Deformation of Elastomer Gels in Equilibrium under Saturated and Unsaturated Conditions," Journal of the Mechanics and Physics of Solids, 59(3), 561-582.

[43] Bouklas N, Huang R (2012) "Swelling Kinetics of Polymer Gels: Comparison of Linear and Nonlinear Theories," Soft Matter, 8(31), 8194-8203.

[44] Durning C, Morman Jr K (1993) "Nonlinear Swelling of Polymer Gels," The Journal of Chemical Physics, 98(5), 4275-4293.

[45] Dolbow J, Fried E, Ji H (2004) "Chemically Induced Swelling of Hydrogels," Journal of the Mechanics and Physics of Solids, 52(1), 51-84.

[46] Ji H, Mourad H, Fried E, Dolbow J (2006) "Kinetics of Thermally Induced Swelling of Hydrogels," International Journal of Solids and Structures, 43(7), 1878-1907.

[47] Feynman R, Leighton R, Sands M (1963) The Feynman Lectures on Physics, Addison-Wesley

[48] Duda FP, Souza AC, Fried E (2010) "A Theory for Species Migration in a Finitely Strained Solid with Application to Polymer Network Swelling," Journal of the Mechanics and Physics of Solids, 58(4), 515-529.

[49] Yan H, Jin B (2012) "Influence of Microstructural Parameters on Mechanical Behavior of Polymer Gels," International Journal of Solids and Structures, 49(3), 436-444.

[50] Wu P, van der Giessen E (1995) "On Neck Propagation in Amorphous Glassy Polymers under Plane Strain Tension," International Journal of Plasticity, 11(3), 211-235.

[51] Valanis K, Landel R (1967) "The Strain-Energy Function of a Hyperelastic Material in terms of the Extension Ratios," Journal of Applied Physics, 38(7), 2997-3002.

[52] Boyce MC (1996) "Direct Comparison of the Gent and the Arruda-Boyce Constitutive Models of Rubber Elasticity," Rubber chemistry and technology, 69(5), 781-785.

[53] ASTM-D412-98a (1998), Standard Test Methods for Vulcanized Rubber and Thermoplastic Elastomers: Tension, American Society for Testing and Materials.

[54] ABAQUS (2012), Software version 6.11: Dassault Systems.

[55] ASTM-D575-91 (2007), Standard Test Methods for Rubber Properties in

Compression, American Society for Testing and Materials.

[56] Qamar SZ, Hiddabi SA, Pervez T, Marketz F (2009) "Mechanical Testing and Characterization of a Swelling Elastomer," Journal of Elastomers and Plastics, 41 (5), September 2009, p 415-431

[57] Qamar SZ, Pervez T, Akhtar M, Al-Kharusi MSM (2012) "Design and Manufacture of Swell Packers: Influence of Material Behavior," Materials and Manufacturing Processes, 27 (7), 2012, p 727-732

Chapter 7

Mechanical and Structural Behavior of Swelling Elastomers under Compressive Loading

Sayyad Zahid Qamar, Maaz Akhtar and Tasneem Pervez

The history of science shows that theories are perishable. With every new truth that is revealed we get a better understanding of Nature and our conceptions and views are modified.

Nikola Tesla

Abstract

Swelling elastomers are a new breed of advanced polymers, and found increasing use in drilling of difficult oil and gas wells. It is important to know how an elastomer will behave under a given set of well conditions, especially after the initial quick-swell period. Good design depends on appropriate material selection. Results are presented in this chapter from experimental and numerical studies conducted to analyze how compressive and bulk behavior of actual oilfield elastomers changes due to swelling. Six key attributes of swelling elastomers needed for design improvement and performance analysis of elastomer seals are discussed: four mechanical properties (elastic modulus E, bulk modulus K, shear modulus G, and Poisson's ratio ν), and two polymer structure characteristics (cross-link chain density N_C, and average molecular weight M_C). These parameters were experimentally determined before and after various stages of swelling for two different swelling elastomers being currently used by the regional petroleum industry, in low and high salinity brines. To strengthen the experimental results, and to be able to forecast for other elastomer materials and well conditions, tests were also simulated using the commercial FEM package ABAQUS, using the best available hyperelastic material models.

Keywords: swelling elastomer, compressive loading, mechanical testing, elastomer structure

1. Introduction

As discussed earlier, swelling elastomers are a new breed of advanced polymers [1]. Over the last two decades, they have found increasing use in drilling of difficult oil and gas wells, remediation of damaged wells, and renewal of production from abandoned wells [2–8]. It is important to know whether an elastomer type or a certain seal design will function properly and reliably under a given set of well conditions [9]. Investigation of the swelling and mechanical behavior of these elastomers is essential in resolving application and design issues. When an elastomer seal is positioned in the well, environmental conditions initiate the swelling. The material may continue to swell throughout its life, though very slowly after the initial quick-swell period [10, 11]. Swelling results in changes in its mechanical properties and sealing response. If elastomer seals are put into use without thoroughly studying these changes, seal failure can happen in certain cases, causing loss of time, money, and other resources. Rework is not only time consuming but at times not possible at all. Failures may even cause the production from the well to stop.

One critical aspect of good design is appropriate material selection. This cannot be done without reliable knowledge of how the material will behave when exposed to different loads, temperatures, and other environmental conditions. As swelling elastomers are basically used as sealing elements in petroleum applications, and as seals are mostly under compressive loading, investigation of the mechanical behavior of elastomers under compression is very important. Contemporary elastomer seal design is usually based on laboratory tests and trial-and-error method. Testing under all possible combinations of field parameters (elastomer material, water salinity, oil type, temperature, pressure, etc) in the laboratory is almost impossible, especially for extended swelling periods. Numerical modeling and simulation of seal design, validated by experimental results, may provide answers for possible scenarios, which cannot be attempted experimentally. Numerical simulations can shorten the lead time for development of robust sealing applications in difficult or new oil and gas wells, offering an accurate and cost effective alternative to extensive testing.

Results are presented in this chapter from experimental and numerical studies conducted to analyze how compressive and bulk behavior of actual oilfield elastomers changes due to swelling. Six key attributes of swelling elastomers needed for design improvement and performance analysis of elastomer seals are discussed: four mechanical properties (elastic modulus E, bulk modulus K, shear modulus G, and Poisson's ratio ν), and two polymer structure characteristics (cross-link chain density N_C, and average molecular weight M_C). These parameters were experimentally determined before and after various stages of swelling for two different swelling elastomers being currently used by the regional petroleum industry, in low and high salinity brines. To strengthen the experimental results, and to be able to forecast for other elastomer materials and well conditions, tests were also simulated [12] using the commercial FEM package ABAQUS [13], using the best available hyperelastic material models.

2. Mathematical background

As stated above, this chapter describes the effect of swelling on mechanical properties (macro-level behavior) and polymer structure (micro-level characteristics) of two elastomers under compressive loading. Brief theoretical background and inter-relationships of these quantities are presented in this section.

2.1 Macroscopic behavior: isotropic relationships

The elastic behavior of any isotropic material is completely described by Hooke's law [14]:

$$\varepsilon_x = \frac{1}{E}\left[\sigma_x - \nu\left(\sigma_y + \sigma_z\right)\right]$$

$$\varepsilon_y = \frac{1}{E}\left[\sigma_y - \nu(\sigma_z + \sigma_x)\right]$$

$$\varepsilon_z = \frac{1}{E}\left[\sigma_z - \nu\left(\sigma_x + \sigma_y\right)\right], \tag{1}$$

where σ is the engineering stress and ε is the strain. Being in the elastic region, engineering shear strain (γ) is obviously proportional to shear stress (τ):

$$\gamma_{yz} = \frac{1}{G}\tau_{yz}, \gamma_{zx} = \frac{1}{G}\tau_{zx}, \gamma_{xy} = \frac{1}{G}\tau_{xy} \tag{2}$$

Figure 1 shows a body under a state of pure shear and its Mohr circle representation giving the principal stresses.

$$\sigma_1 = \tau_{xy}, \quad \sigma_2 = -\tau_{xy}, \quad \sigma_3 = 0, \tau_{12} = 0 \tag{3}$$

We can use Hooke's law Eq. (1) to find the principal strains:

$$\varepsilon_1 = \frac{\tau_{xy}(1+\nu)}{E}, \varepsilon_2 = -\frac{\tau_{xy}(1+\nu)}{E}. \tag{4}$$

Shear strain γ_{xy} can also be expressed in terms of the principal strains:

$$\gamma_{xy} = \varepsilon_1 - \varepsilon_2 = \frac{2\tau_{xy}(1+\nu)}{E} \tag{5}$$

Comparing Eqs. (2) and (5), shear modulus becomes:

$$G = \frac{E}{2(1+\nu)}. \tag{6}$$

Bulk modulus (K) is related to volume dilatation (volumetric strain) $\delta = \Delta V/V$ and hydrostatic pressure$p = \frac{1}{3}\left(\sigma_x + \sigma_y + \sigma_z\right)$:

$$\delta = -\frac{p}{K}. \tag{7}$$

For a body undergoing linear strains, it can be easily shown that [15].

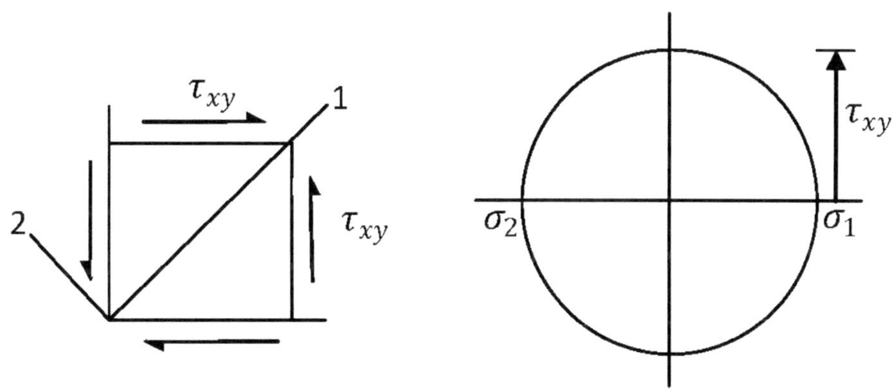

Figure 1.
A body under state of pure shear and its Mohr circle representation.

$$\frac{dV}{V} = d\varepsilon_x + d\varepsilon_y + d\varepsilon_z. \tag{8}$$

Combining Eqs. (7) and (8), volume dilatation can be expressed as

$$\delta = \varepsilon_x + \varepsilon_y + \varepsilon_z. \tag{9}$$

Comparing this result with Eq. (1), and simplifying, we get

$$\delta = \left(\frac{1-2\nu}{E}\right)\left(\sigma_x + \sigma_y + \sigma_z\right). \tag{10}$$

Recalling the definition of hydrostatic pressure p, and comparing with Eq. (7), the relationship between bulk modulus and Poisson's ratio comes out to be

$$K = \frac{E}{3(1-2\nu)}. \tag{11}$$

Once the values of E and K are established experimentally, shear modulus (G) and Poisson's ratio (ν) can be determined using the isotropic relations Eqs. (6) and (11).

2.2 Structural behavior

Chemical and physical structure of a polymer is modified at different stages of swelling. The resulting changes in structural behavior affect the performance of swelling elastomer seals. It is important, therefore, to understand the factors that can affect swelling elastomer structure and, hence, its properties. Current section discusses the method of determining major structural properties.

2.2.1 Chain density

For any rubber-like material, network strands are formed both by cross-linking and by molecular entanglement. As the degree of cross-linking increases, the cross-link chain density increases, making the elastomer more rigid, thus giving higher values of stiffness and shear modulus G [16]. Chain density can therefore be used as a measure of the swelling capability of an elastomer. For elastomers subjected to small strains, shear modulus is directly related to the number of cross-link chains per unit volume N_c [17]:

$$G = N_c kT, \tag{12}$$

where k is the Boltzmann's constant, and T is the temperature in Kelvins.

2.2.2 Number average chain molecular weight

Length of the network chain decreases due to an increase in the degree of cross-linking. In any polymer, chain density is closely related to the average molecular weight, and this property can also be used to explain the mechanics of swelling [17]. Multiplying and dividing Eq. (12) by Avogadro's number N_A, and recalling that $R = N_A k$, we can obtain

$$G = \frac{N_c RT}{N_A}. \tag{13}$$

If n is the number of chains, Avogadro's number can also be written as

$$N_A = nM_c/m. \tag{14}$$

Combining the definition of chain density ($N_c = n/V$) with Eq. (14) yields

$$N_c/N_A = \rho/M_c. \tag{15}$$

Comparing this result with Eq. (13), we can get the expression for the number average chain molecular weight

$$M_c = \rho RT/G. \tag{16}$$

Once the value of shear modulus G is determined, these structural properties of the elastomer (N_c and M_C) can be easily evaluated using Eqs. (12) and (16).

3. Experimental work

Various researchers have investigated the behavior of rubber-like materials under compression [16], but not for swelling elastomers. The main objective of experiments described here is to obtain swelling behavior, stress–strain, and pressure-volumetric strain curves. Curve fitting on the linear portions of these graphs determines the values of elastic and bulk moduli. These curves are also used to extract material coefficients for the hyperelastic models used for numerical simulations.

Samples of two different water-swelling elastomers were allowed to swell for a one month period in saline water of 6,000 ppm (0.6%) and 120,000 ppm (12%) concentrations at 50°C; actual field conditions in regional oil-wells. Samples were placed in sealable glass jars containing the swelling medium and identified by mnemonic code names. Jars were placed inside servo-controlled ovens maintained at 50°C temperature throughout the thirty days test period. Compression and bulk tests were carried out on disc samples before swelling and after 1, 2, 4, 7, 10, 16, 23, and 30 days of swelling. To ensure repeatability and consistency, all reported values are average of readings from three samples.

3.1 Swelling tests

Cutting of disc samples and preparation of salt solutions has been described in detail in Chapter 3. Procedure for swelling tests is given below.

3.1.1 Test procedure

Samples were placed in sealable glass jars containing proper swelling medium and identified by mnemonic codenames. Jars were placed inside servo-controlled ovens maintained at 50°C temperature throughout the 30-day test period. Volume, thickness, mass, and hardness of each sample was measured before swelling and after different days of swelling. Test setup has been described in Chapter 3. Mass was measured using a digital balance. Volume measurements were done by a specially constructed (in-house) graduated beaker-cylinder arrangement based on the

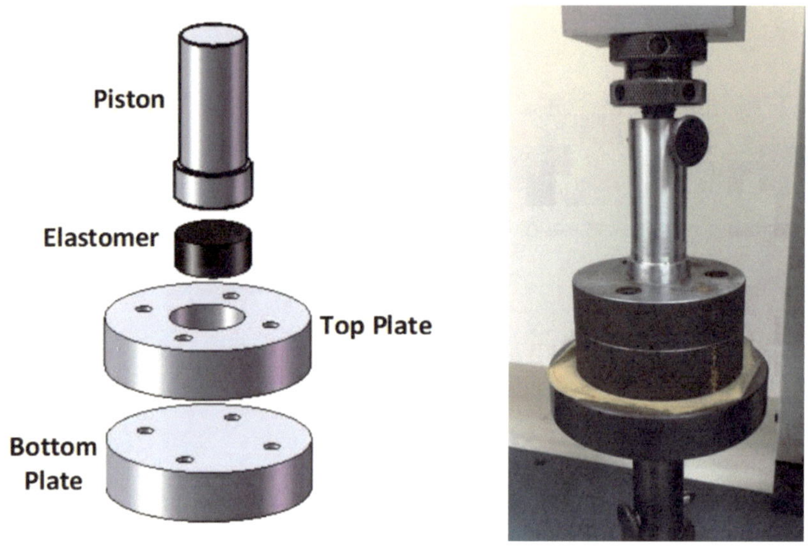

Figure 2.
Designed and fabricated test rig for bulk experiments.

displacement method. Thickness measurements were carried out using digital Vernier calipers. Hardness readings were taken using a scale-A durometer.

3.2 Compression testing

Compression tests are performed in accordance with ASTM-D575 standard test method [18] as described in detail in Chapter 6.

3.3 Bulk testing

No standard is available for bulk testing of elastomers. A special test rig was therefore designed and fabricated in-house for bulk experiments; **Figure 2**. The rig was designed in such a way that under compressive loading the specimen is constrained to move only in the axial direction and radial expansion is not allowed. To match with compression testing, load is applied at the rate of 12 mm/min until the specimen fails. Force-deformation results are recorded and later converted into pressure-volumetric strain data.

4. Numerical investigation

Just as in experiments, simulations are run before swelling and after every swelling period. Using the commercial finite element package ABAQUS, swelling elastomer specimen is modeled using 8-noded linear brick element with reduced integration (C3D8R). As mentioned in Chapter 3, Ogden hyper-elastic material model with second strain energy potential [19] is used for all simulations since it gives the best results [20]. Coefficients for the Ogden model are extracted from the experimental stress–strain data.

4.1 Simulation of compression test

Simulations for compression experiments are discussed in detail in Chapter 3. Stress–strain curves from these simulations are used here to determine the values of

Young's modulus. Simulations are conducted before swelling and after each swelling period (total one month period) under both salinity conditions, and for both materials. Specimen geometry before and after deformations is shown in **Figure 3**.

4.2 Simulation of bulk test

For bulk testing, elastomer disc is still under compression but restricted from any transverse deformations. Only the top surface can move downward (z-axis) under the compressive load as shown in the test setup in **Figure 2**. Simulation results are used to plot graphs of pressure against volumetric strain, for determination of bulk modulus through curve fitting. Deformed and undeformed specimens from bulk test simulation are shown in **Figure 4**.

5. Results and discussion: experimental

Results of the experimental investigation (mechanical and structural response of the elastomer to swelling) are discussed in this section. Variations in properties (volume, thickness, hardness, and density) due to swelling are discussed first. Later, results from compression and bulk experiments are presented and analyzed.

5.1 Swelling behavior

Effect of swelling on volume, thickness, hardness, and density of the two types of elastomer samples under different salinities is discussed below.

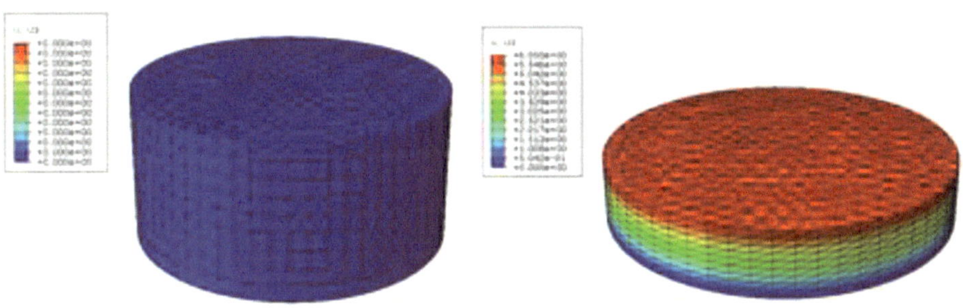

Figure 3.
Finite element model of compression test specimen; undeformed and deformed states.

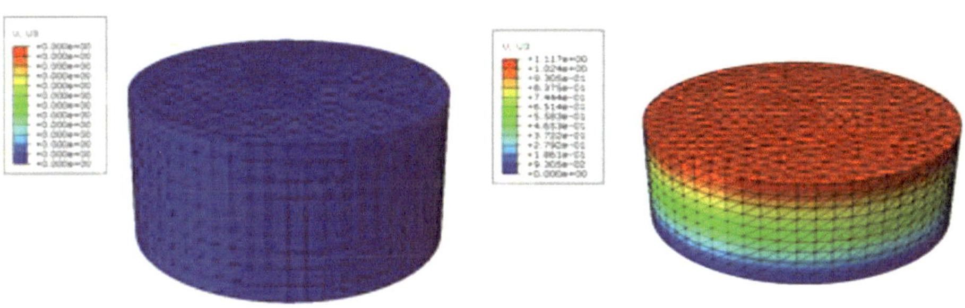

Figure 4.
Finite element model of bulk test specimen; undeformed and deformed states.

5.1.1 Volume change

Figure 5 shows the change in volume against days of swelling in 6000 ppm (0.6%) and 120,000 ppm (12%) brine solutions at 50°C for materials *A* and *B*. Both elastomers are of the fast-swell type, so increase in volume is very quick in the beginning, which can be observed in the first five days (reaching to almost one-third of total volume increase). More gradual increase was observed for the remaining one month period. As observed in earlier studies, swelling in lower-concentration salt-solution is higher than in higher-concentration brine due to existence of higher chemical potential gradient.

5.1.2 Thickness change

Figure 6 shows thickness change against swelling time for both materials. Fast swelling nature of the elastomer (materials *A* and *B*) results in more thickness increase in the beginning, and slower increase later on. Again, as expected, thickness swelling is higher for lower salt concentration.

5.1.3 Hardness change

Hardness keeps on decreasing as the elastomer swells more. **Figure 7** summarizes the change in hardness as samples for both materials swell in 0.6% and 12%

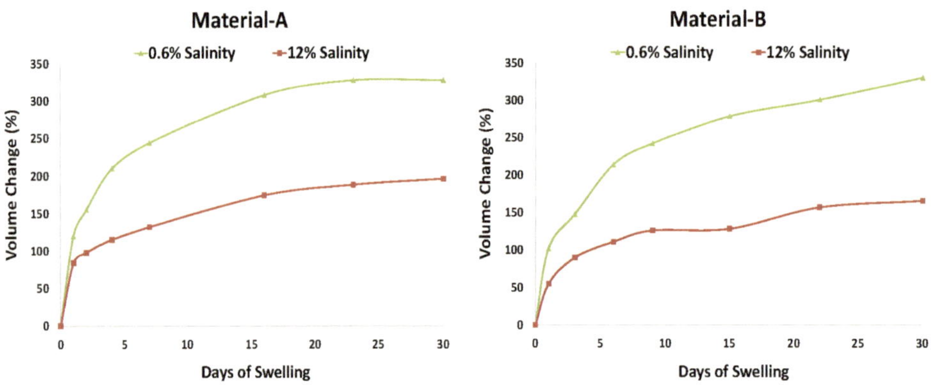

Figure 5.
Volume change in both materials in 0.6% and 12% salinity at 50°C.

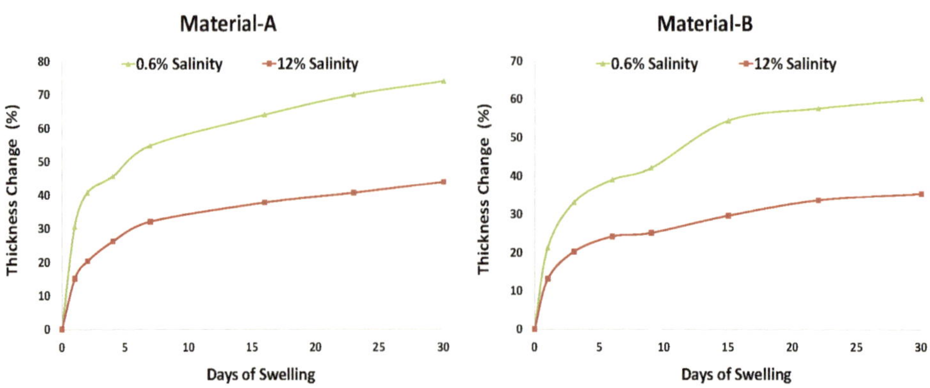

Figure 6.
Thickness change in both materials in 0.6% and 12% salinity at 50°C.

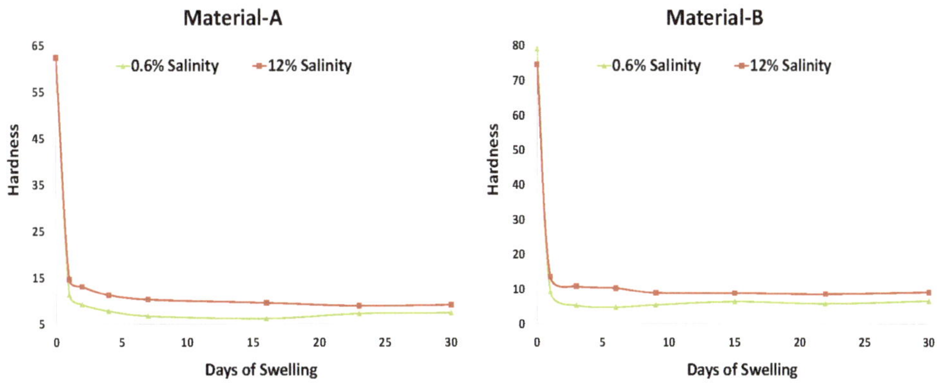

Figure 7.
Hardness change in both materials for 0.6% and 12% salinity at 50°C.

salt solutions, at 50°C. It is interesting to observe that within the first week of swelling, hardness drastically drops down, and then remains almost constant. Material *B* is harder in nature (unswelled hardness of about 75 shore-A as compared to 62 shore-A for material *A*). Hardness drop after swelling is more under 0.6% salinity solution for both materials.

In all of the above graphs, the overall trend is the same; very high swelling rate in the beginning, then a slow gradual change. This leads to a sharp initial drop in hardness, followed by almost no change for longer swelling periods. The slight fluctuating pattern may have roots in the nature of the swelling process [11, 21, 22]. Salt is one of the constituents of any swelling elastomer. When exposed to saline water, some salt can enter into the elastomer material (together with absorbed water), while some salt can break away from the elastomer and go into the salt solution. Rather than a consistent increase, this two-way salt transport cause slight fluctuations in the amount of swelling. Another important factor in swelling of such elastomers is the density of cross-link chains. Salt addition and breakaway can produce breaking and later re-forming of cross-link chains, causing increase or decrease of swelling at a particular time. As both these mechanisms are dependent on salt transport, fluctuation is more significant in higher concentration brine. After some time, a sort of equilibrium is reached, inflow and outflow of salt almost balancing out, and swelling starts to show a near steady-state behavior.

5.1.4 Density change

Density drops down considerably in the first few days for both materials and salinities, as can be seen in **Figure 8**. With more swelling, it increases a little before gradually dropping down to almost the same value as at the end of the first week. Due to the two-way water and salt transport during swelling of these elastomers, both volume and mass change at different rates; thus the slightly zigzag pattern of density variation.

5.2 Behavior under compressive loading

As an example, **Figure 9** shows 3-sample compressive stress–strain graphs for materials *A* and *B* in high salinity solution after four days of swelling. It can be seen that stress–strain behavior is non-linear in nature, as expected. It can also be observed that material *B* is more rigid than material *A*; thus the higher stresses and lower strains.

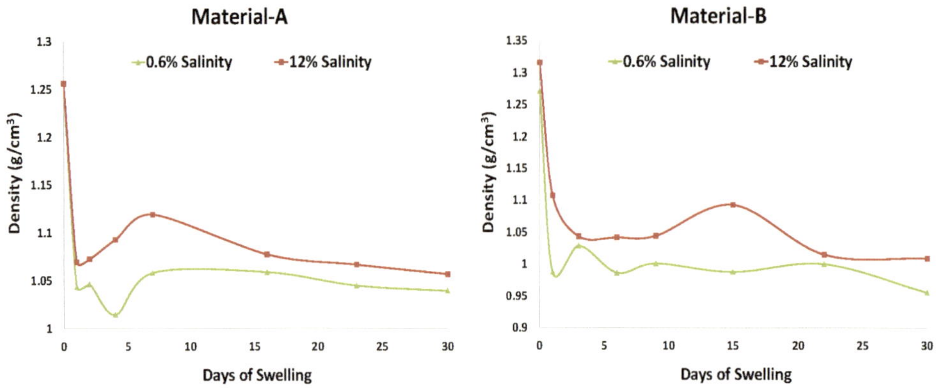

Figure 8.
Density variation in both materials for 0.6% and 12% salinity at 50°C.

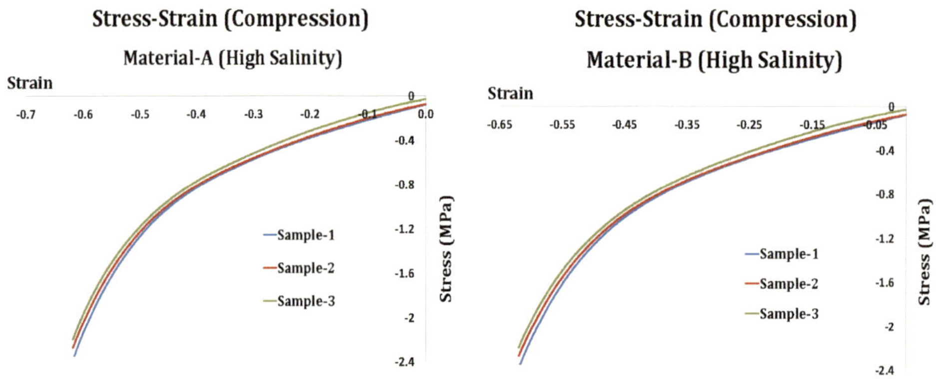

Figure 9.
Stress–strain graphs for both materials at 4 days of swelling; 3 samples.

Three-sample pressure-volumetric strain graphs for both elastomer types under high salinity after 23 days of swelling are shown in **Figure 10**. Though the overall behavior is still nonlinear, as expected for rubbers and elastomers, there is a significant linear portion. Bulk modulus is determined by fitting a straight line to the linear portion of these pressure-volumetric strain curves.

5.2.1 Mechanical and structural properties

At small strains (up to 10%), elastomers exhibit almost a linear stress–strain relationship and behave like other elastic materials [23]. Hence, for the extraction of Young's modulus, curve fitting is done on the initial 10% portion of the compressive stress–strain curve. For the determination of bulk modulus, some initial pressure-strain data is discarded before curve fitting. Elastomer disc does not touch the walls of the test cylinder at all points at the beginning of the test. As pressure is applied and the disc expands, it starts to fill this gap. Actual bulk test starts at the end of this gap-filling stage. Samples of linear curve fitting for extraction of elastic and bulk moduli (E and K) are shown in **Figure 11**. A very high goodness of fit of more than 99% is quite reassuring. Based on these values of E and K, shear modulus and Poisson's ratio are calculated using isotropic relations for elastic materials Eqs. (6) and (11). Chain density and average molecular weight are determined through polymer structure relations Eqs. (12) and (16).

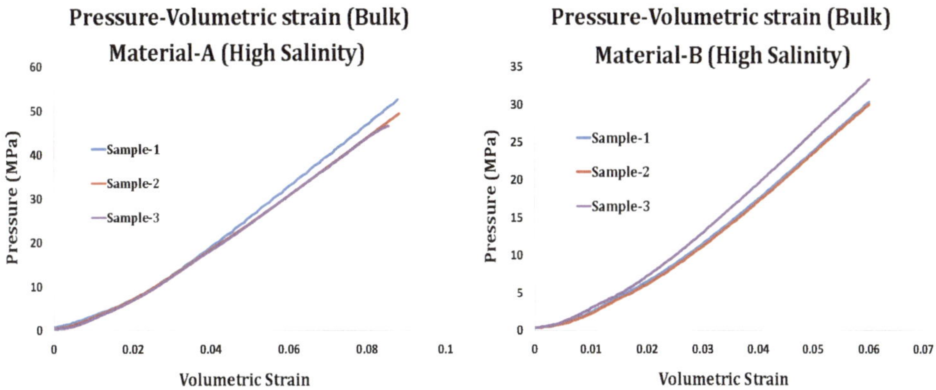

Figure 10.
Three-sample pressure-strain (volumetric) curves for both materials for bulk test; 12% salinity; 23 days of swelling.

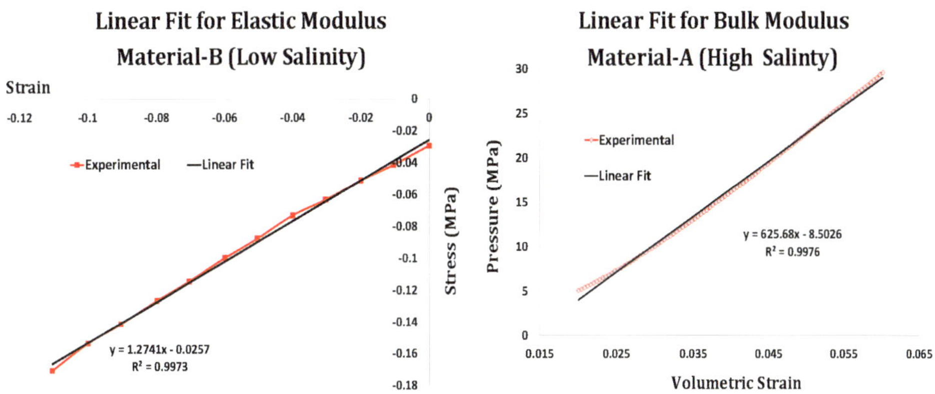

Figure 11.
Linear curve fitting to experimental stress–strain data for determination of Young's modulus (left), and pressure-volumetric strain for determination of bulk modulus (right).

5.2.2 Effect of swelling time

Compressive stress–strain behavior at different stages during the one month swelling period for materials *A* and *B* in high salinity brine is shown in **Figure 12**. Within a day of swelling, stress values drop by about 80–90% in both high and low salinities. Stress–strain curves after that are close to each other, the difference between curves for day-1 and day-30 not being very high. Lower stresses can be observed in low salinity solution for both materials, as expected; this is because high swelling is taking place in low salinity medium (as reported above), leading to softer samples. Also, strains after longer swelling periods are much larger. With more swelling, elastomer becomes softer, giving large strains at small loads, resulting in the trend shown in **Figures 9** and **10**.

Pressure-volumetric strain curves for one month period for both materials in low salinity are shown in **Figure 13**. A trend similar to the compression curves can be seen in these pressure-strain graphs also; larger drop in bulk modulus during the first few days of swelling, then more gradual. Reasons are the same as discussed above.

5.2.3 Variation of elastic Modulus

Variation of Young's modulus (*E*) under compression for materials *A* and *B* under both salinities is shown in **Figure 14**. *E*-value decreases sharply in the first

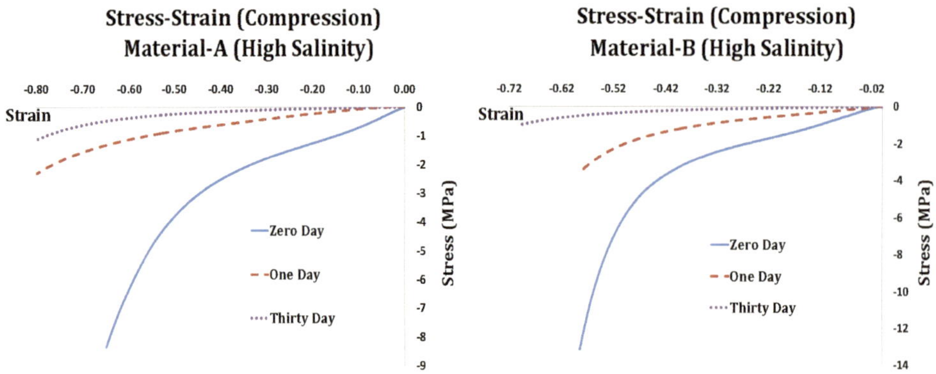

Figure 12.
Compressive stress–strain curves for both materials in high salinity for one month period.

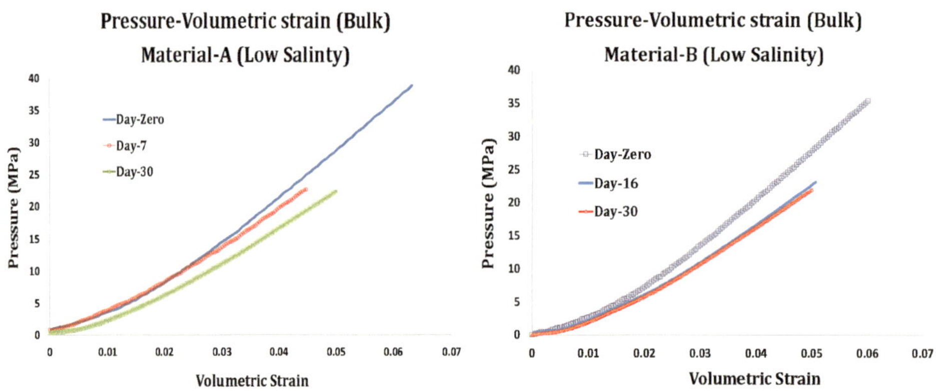

Figure 13.
Pressure-volumetric strain behavior before swelling and after 7 and 30 days of swelling in low salinity brine.

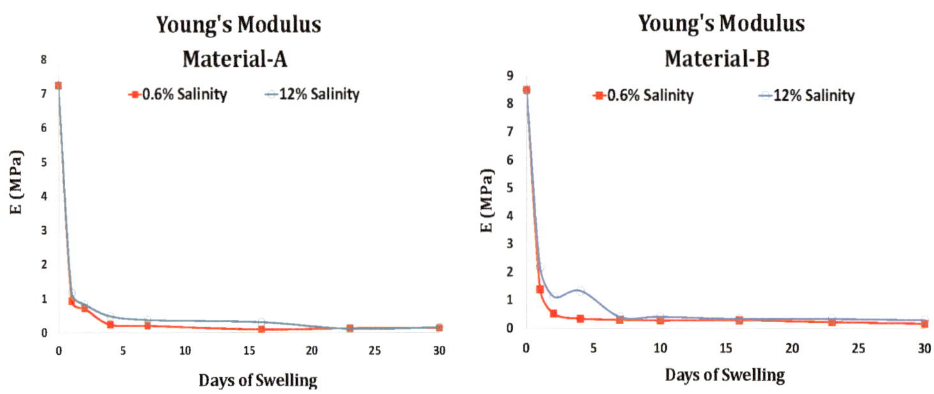

Figure 14.
Experimental results for Young's modulus for both materials.

few days (by almost 90%) and then becomes almost constant. This overall decreasing trend and slight fluctuation is due to the nature of swelling elastomers, as explained above. Modulus values for 12% salinity are slightly higher, as these samples are somewhat stiffer than the lower-salinity ones.

5.2.4 Variation of bulk Modulus

Variation of bulk modulus (K) over the 30-day swelling period for materials A and B under 0.6% and 12% salinities is shown in **Figure 15**. There is a notable fluctuation in the variation of K-value especially for material B; a sharp initial decrease, then a small increase, followed by a more gradual decrease to almost a constant value at the end of the one-month swelling period. Apart from the reasons for fluctuation discussed above, significant oscillation in K-value may also be rationalized from another perspective. During the compression test, sample elastomer disc is compressed in the axial direction, but is free to expand in the radial direction. However, in the bulk test, the disc specimen is constrained inside a cylinder so that there is uniaxial compression and no radial expansion. In effect, this violation of volume constancy means that the elastomer behavior changes from incompressible (compression test) to compressible (bulk test). This may lead to notable deviation from regular trends. As for salinity, the behavior is similar to that under compression; higher salinity gives larger values of bulk modulus.

5.2.5 Variation of Poisson's ratio

Variations of Poisson's ratio (ν) for both materials and both salinities exhibits a mirror behavior in comparison with elastic modulus, as shown in **Figure 16**. Poisson's ratio increases sharply in the first few days, and then becomes almost constant for the remaining swelling period. Again, lower salinity is giving higher values, as expected. For material B, there is some fluctuation during days 2 to 7 under 12% salinity. This variation pattern is similar to that of Young's modulus (**Figure 14**). As ν is dependent on E (isotropic relation), this fluctuation is expected. Moreover, it is interesting to note that ν approaches the limiting value of 0.5 within the first 10 days of swelling. This means that the assumption of incompressibility used in most analytical and numerical models of rubber-like materials is also justified for swelling elastomers, given a reasonably large swelling period.

5.2.6 Variation of shear Modulus

Variation of shear modulus (G) for both materials under low and high salinities is shown in **Figure 17**. Just like Young's modulus, value of shear modulus drops by more than 90% in the first 5 to 6 days, and then remains practically constant during the rest of the swelling period. Except for the first few days, both salinities show the

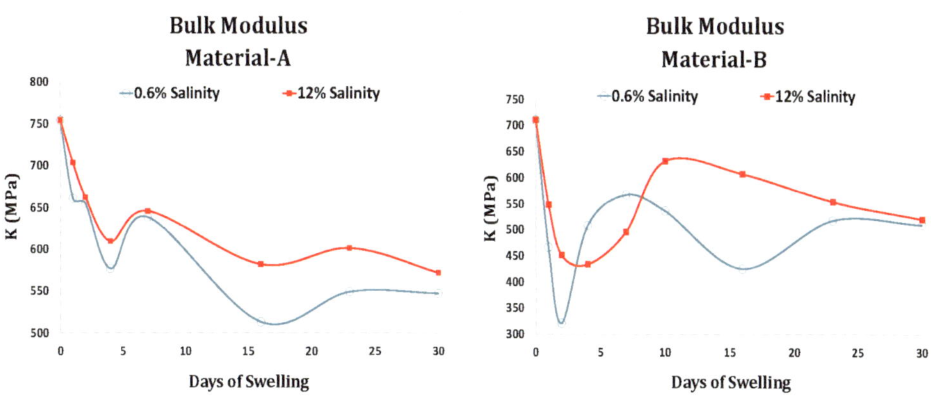

Figure 15.
Experimental results for bulk modulus for both materials.

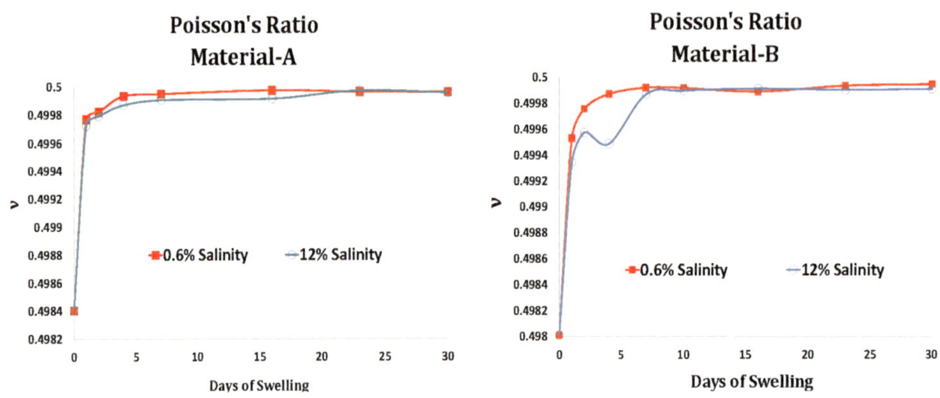

Figure 16.
Experimental results for Poison's ratio for both materials.

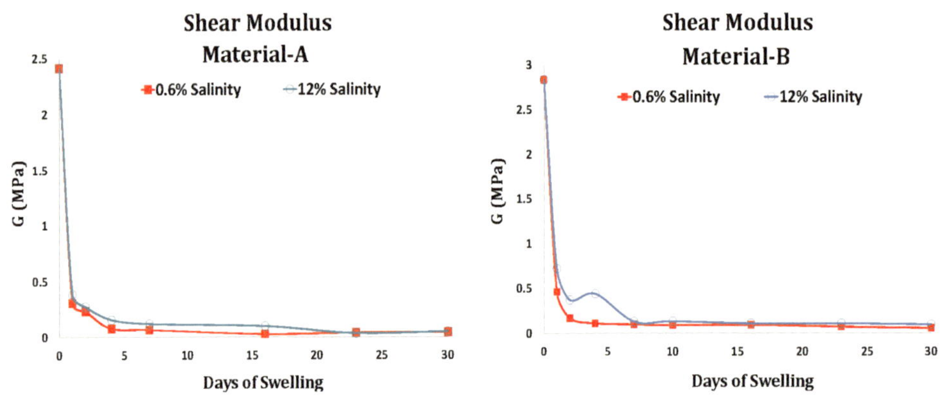

Figure 17.
Experimental results for shear modulus for both materials.

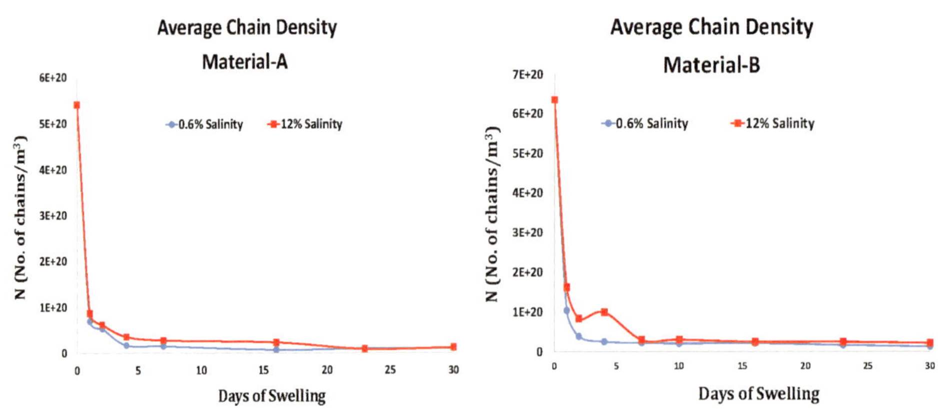

Figure 18.
Experimental results for average chain density for both materials.

same values. Some fluctuations for material *B* under 12% salinity can be observed during days 2 to 7. As *G* is related isotropically to *E*, it exhibits similar behavior as observed in Young's modulus.

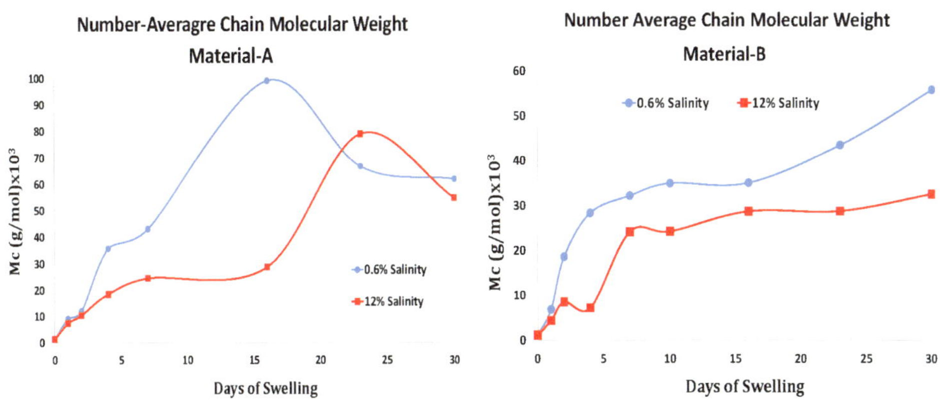

Figure 19.
Experimental results for number-average chain molecular weight for both materials.

5.2.7 Variation of chain density

Variation in cross-link chain density (N) against amount of swelling is plotted in **Figure 18** for materials A and B in both salinities. Chain density drops drastically during the first few days, and then becomes nearly steady. During days 2 to 7, values are higher for material B in higher-salinity solution, otherwise response under both salinities is almost the same. As explained above, decrease in number of chains or cross-linking density occurs due to swelling, which makes the elastomer softer [11, 21, 22], thereby contributing to a decrease in stiffness, elastic modulus, and rigidity (shear modulus). This change in polymer structure due to swelling is a direct explanation of the variation pattern observed in the mechanical properties.

5.2.8 Variation of molecular weight

Figure 19 shows the variation in number-average chain molecular weight (M_C) against swelling time over the 30-day test period for materials A and B in both salinities. There is an overall increasing behavior with higher swelling amount, but with a somewhat fluctuating pattern, material A exhibiting higher fluctuations. The various reasons discussed above (including two-way transport of salt, forming and breaking of cross-link chains, etc) are equally applicable in explanation of the observed variation in average molecular weight. On top of that, M_c is related to both density and shear modulus, so it is exhibiting the combined effect of two fluctuations.

Slight fluctuations during the first week of swelling in the values of material and structural properties studied are due to this two-pronged variation in swelling amount. Compared to bulk tests, fluctuation is almost insignificant in compression tests due to the simple uniaxial nature of the load applied. ***As significant variation in material properties occurs only in the first 10 days, these values can serve as reasonable design properties for future applications, removing the need for material testing beyond 10 days of swelling***.

6. Results and discussion: numerical

Stress–strain curves under compressive loading, and pressure-strain curves for bulk testing are also produced through numerical simulation, using Ogden hyperelastic material model with second-degree strain energy function (N = 2). Graphs for variation of mechanical properties against swelling are generated from the numerical results. This work has been done for both materials under both salinities for the entire one-month swelling period. However, only a few representative graphs are presented below for comparison of experimental and numerical results.

6.1 Behavior under compressive loading

Stress–strain plots in the 10% strain region for experimental and numerical results of compression test for material *B* under 12% salinity are shown in **Figure 20**, before swelling, and after one day of swelling. Variation of pressure against volumetric strain (experimental and numerical) from bulk test results of material *B* under 0.6% salinity is given in **Figure 21**. Though variation trends are roughly the same, there is considerable difference between experimental and numerical curves in general.

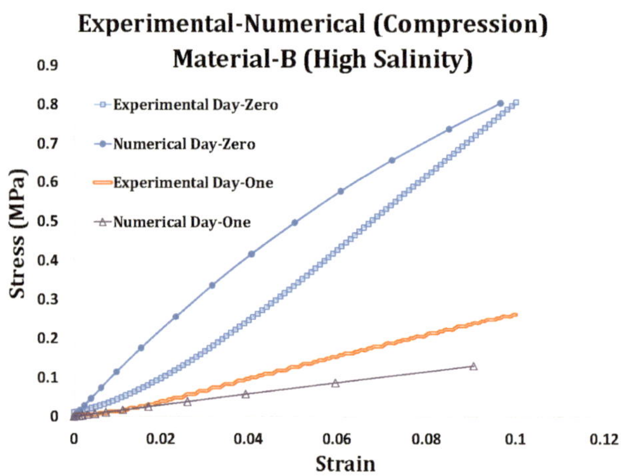

Figure 20.
Comparison of experimental and numerical (Ogden; N = 2) results for compression test for both materials, before swelling (day zero) and after one-day of swelling.

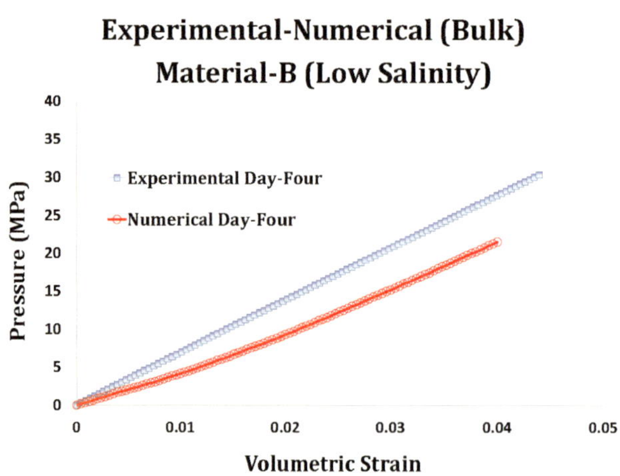

Figure 21.
Comparison of experimental and numerical (Ogden; N = 2) results for bulk test for material B *in low salinity after 4 days of swelling.*

6.1.1 Mechanical properties

Variation of Young's modulus, Poisson's ratio, and shear modulus for the two materials under 0.6% and 12% salinity over the 30-day swelling period is shown in **Figures 22–24**. A reasonable match between experimental and simulation results can be observed. There are notable differences during days 2 to 7, especially in the case of higher salinity (12%), but the curves come close to each other upon further swelling.

Comparison between experimental and numerical results for bulk tests is given in **Figures 25** and **26**, for materials *A* and *B* respectively. Numerical results are seriously different from experimental ones in all cases over the 30-day swelling period. Even in the experimental results, there is serious fluctuation in bulk modulus, as observed and discussed earlier in Section 4. On top of that, it is well known that the available hyperelastic models are based on shear deformation theory, and do not include the effect of bulk deformation. This would obviously imply a notable difference in experimental and simulation results of bulk behavior.

6.1.2 Note

As discussed in detail in Chapter 6, Ogden model (N = 2) is the best one among the available hyperelastic models for representation of swelling elastomers.

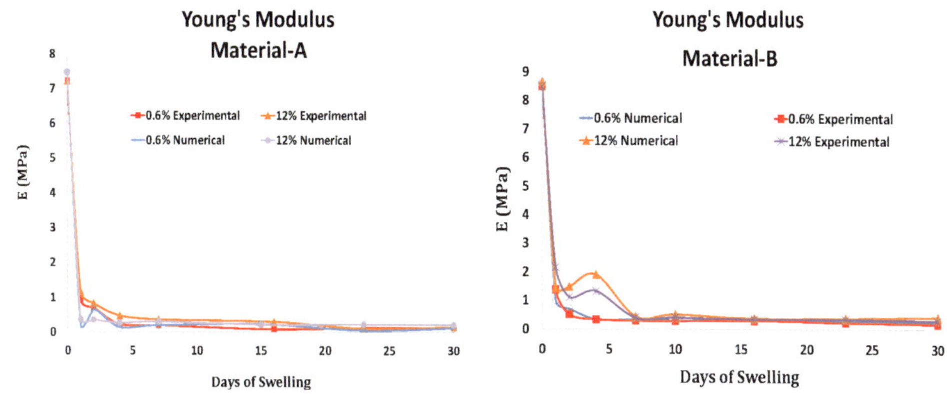

Figure 22.
Comparison of experimental and numerical results for variation of Young's modulus.

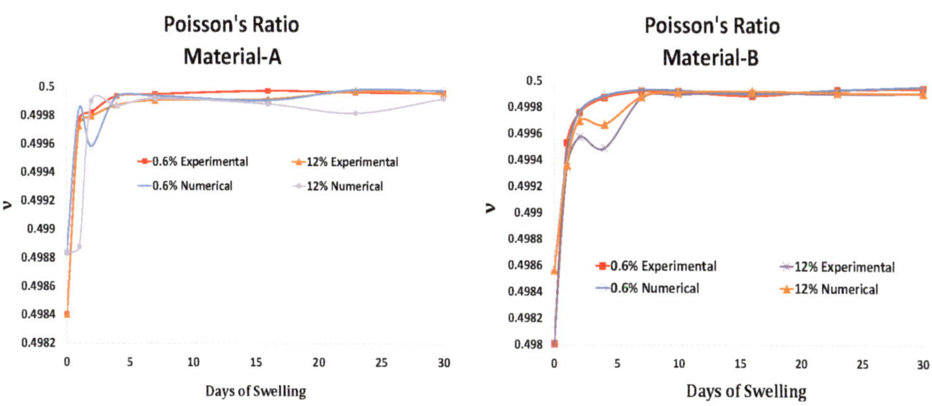

Figure 23.
Comparison of experimental and numerical results for variation of Poisson's ratio.

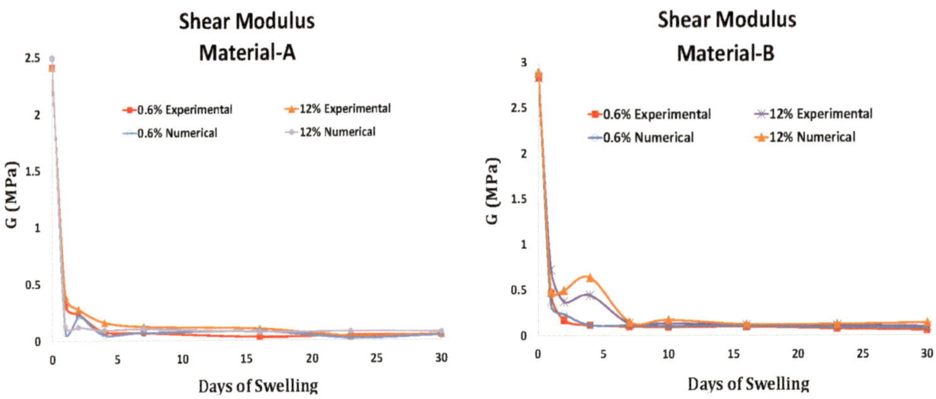

Figure 24.
Comparison of experimental and numerical results for variation of shear modulus.

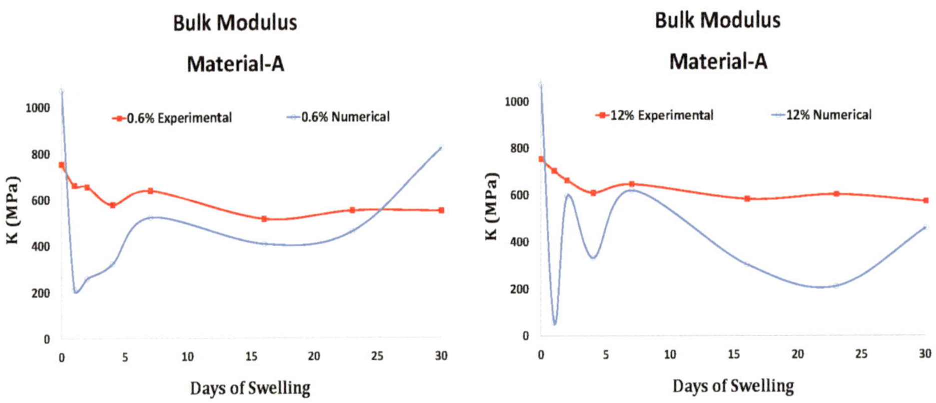

Figure 25.
Comparison of experimental and numerical results for variation of bulk modulus for material A.

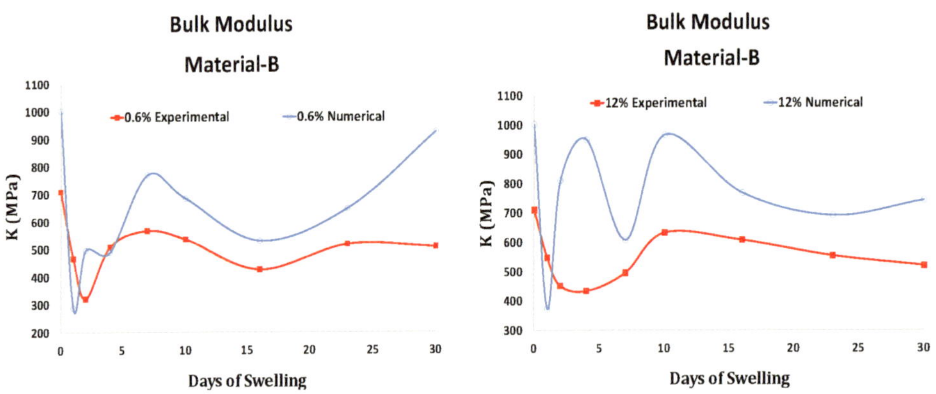

Figure 26.
Comparison of experimental and numerical results for variation of bulk modulus for material B.

However, as most of the graphs in this section show, even the Ogden model does not yield results that follow experimental ones closely. In the absence of any other model, this may be the only choice for numerical simulation, but it is not good enough. ***This highlights the need for a new material model that can more closely***

capture the actual behavior of swelling elastomers. As described later in Chapter 11, this new model should include continuum mechanics, diffusion, and thermodynamics of mixing to account for large deformations in the case of swelling.

7. Conclusions

Experimental and numerical investigation of changes in compressive and bulk behavior of two water-swelling elastomers has been presented. Tests were carried out on standard compression and bulk samples (following ASTM standards) before swelling and after different swelling periods. Elastic modulus and bulk modulus were experimentally determined under different swelling conditions. Shear modulus and Poisson's ratio were estimated using isotropic relations. Cross-link chain density and number average molecular weight were obtained using predictive equations of polymer structure. Mechanical testing was also modeled and simulated using the nonlinear finite element package ABAQUS, using Ogden hyperelastic model with second strain energy potential. Simulation results are somewhat close to experimental ones in most of the cases. However, there are significant differences in various other situations. This strengthens the need for a new material model to more effectively represent the behavior of swelling elastomers.

Author details

Sayyad Zahid Qamar[1]*, Maaz Akhtar[2] and Tasneem Pervez[1]

1 Mechanical and Industrial Engineering Department, Sultan Qaboos University, Muscat, Oman

2 Mechanical Engineering Department, N.E.D. University of Engineering and Technology, Karachi, Pakistan

*Address all correspondence to: sayyad@squ.edu.om

References

[1] Polgar LM, Fallani F, Cuijpers J, Raffa P, Broekhuis AA, van Duin M, Picchioni F (2017) "Water-Swellable Elastomers: Synthesis, Properties and Applications," Reviews in Chemical Engineering, 35 (1), p 45-72

[2] Al-Douseri K, Barnes C, Young D, Smith P (2010) "Swellable Packers Provide a Brown Field Water Management Solution in Open and Cased Hole, SPE Oil and Gas India Conference and Exhibition (OGIC), Mumbai, India.

[3] Ekpe J, Kompantsev A, Al-Thuwaini J, Belousov Y, Ashoor A, Abdul-Aziz ME (2011) "First Multi-Stage Swell Packers Open Hole Completion in Rub Al-Khali Empty Desert: Case Study," Paper # SPE 142523, SPE Middle East Unconventional Gas Conference and Exhibition, Muscat, Oman

[4] Kennedy GP, Lawless AR, Shaikh AK, Alabi IB (2005) "The Use of Swell Packers as a Replacement and Alternative to Cementing," Paper # SPE 95713, SPE Annual Technical Conference and Exhibition, Dallas, Texas

[5] Keshka A, Elbarbay A, Menasria C, Smith P (2007) "Practical Uses of Swellable Packer Technology to Reduce Water Cut: Case Studies from the Middle East and other Areas," Paper # SPE 108613, Offshore Europe, Scotland, UK

[6] Kleverlaan M, van Noort RH, Jones I (2005) "Deployment of Swelling Elastomer Packers in Shell E&P," Paper # SPE/IADC 92346, SPE/IADC Drilling Conference, Amsterdam, Netherlands

[7] Chen X, Bartos J, Salem H, Zonoz R (2016) "Elastomers for High Pressure Low Temperature HPLT Sealing," Paper # IDOTC-27227, Offshore Technology Conference, 2-5 May 2016, Houston, Texas, USA

[8] Ojha K, Saxena A, Pathak AK (2014) "Underbalanced Drilling and Its Advancements: An Overview," Journal of Petroleum Engineering and Technology, Vol 4, No 2

[9] Patel H, Salehi S, Ahmed R, Teodoriu C (2019) "Review of Elastomer Seal Assemblies in Oil and Gas Wells: Performance Evaluation, Failure Mechanisms, and Gaps in Industry Standards," Journal of Petroleum Science and Engineering, 179 (2019), p 1046-1062

[10] Al-Yami AS, Nasr-El-Din HA, Awang MZB, Al-Humaidi AS, Al-Arfaj MK (2008) "Swelling Packers; Lab Testing and Field Application," Paper # IPTC 11997, International Petroleum Technology Conference. Kuala Lumpur, Malaysia

[11] Qamar SZ, Al-Hiddabi SA, Pervez T, Marketz F (2009) "Mechanical Testing and Characterization of a Swelling Elastomer," Journal of Elastomers and Plastics, 41 (5), p 415-431

[12] Qamar SZ, Akhtar M, Pervez T, Al-Kharusi MSM (2013) "Mechanical and Structural Behavior of a Swelling Elastomer under Compressive Loading," Materials and Design, 45 (March 2013), p 487–496

[13] ABAQUS (2012) Software version 6.11, Dassault systems

[14] Hosford WF (2010) Mechanical Behavior of Materials, Cambridge University Press

[15] Hibbeler RC (2016), Mechanics of Materials, 10th edition, Pearson

[16] Gent AN (2012) Engineering with Rubbers: How to Design Rubber Components, Carl Hanser Verlag Publishers, Munich, Germany

[17] Treloar LRG (2005) The Physics of Rubber Elasticity, Oxford University Press, Oxford

[18] ASTM D575–91 (2007) "Standard Test Methods for Rubber Properties in Compression," American Society for Testing and Materials

[19] Ogden R (1972) "Large Deformation Isotropic Elasticity — On the Correlation of Theory and Experiment for Incompressible Rubberlike Solids," Proceedings of the Royal Society of London. A. Mathematical and Physical Sciences, 326(1567), 565-584.

[20] Akhtar M, Qamar SZ, Pervez T, Khan R, Al-Kharusi MSM (2012)" Elastomer Seals in Cold Expansion of Petroleum Tubulars: Comparison of Material Models," *Materials and Manufacturing Processes*, 27(7), p 715-720

[21] Qamar SZ, Pervez T, Akhtar M (2016) "Performance Evaluation of Water-Swelling and Oil-Swelling Elastomers," Journal of Elastomers and Plastics, 48 (6) 2016, p 535-545

[22] Qamar SZ, Pervez T, Akhtar M, Al-Kharusi MSM (2012) "Design and Manufacture of Swell Packers: Influence of Material Behavior," Materials and Manufacturing Processes, 27 (7), p 721-726

[23] Gent A (1996) "A New Constitutive Relation for Rubber," Rubber Chemistry and Technology, 69(1), 59-61.

Chapter 8

Numerical Investigation of Elastomer Seal Performance

Sayyad Zahid Qamar, Maaz Akhtar and Tasneem Pervez

The important thing in science is not so much to obtain new facts as to discover new ways of thinking about them.

Sir William Bragg

Abstract

Analytical models for swelling of rubberlike materials are difficult to formulate, and restricted in actual application due to their need for simplifying assumptions. Tests conducted on laboratory size samples of swelling elastomers cannot reproduce actual oil well conditions, and cannot cover all possible variations of testing parameters. However, these laboratory tests do provide useful information about material response of swellable elastomers in various conditions, serving as a basis for analytical and numerical modeling. Properly developed and robust numerical models can be used to predict near-actual performance of elastomeric seals. The current chapter describes the use of numerical (finite element) simulation to investigate swelling elastomer seal behavior in downhole petroleum applications. Variations in sealing (contact) pressure are studied for seal length, seal thickness, compression ratio, water salinity, swelling time, and type of well completion (open-hole or cased-hole). Month-long swelling experiments on samples of two actual elastomers (Chapters 3 and 7) provide input to the numerical model in terms of real material and deformation data. On the basis of these results, petroleum engineers can make informed decisions about the selection of elastomer material and seal geometry appropriate for the well type and conditions encountered. Application developers and researchers can also find this investigation useful in performance analysis and design of swelling elastomer seals.

Keywords: swelling elastomer seal, numerical simulation, seal contact pressure, different sealing parameters

1. Introduction

Effect of swelling on mechanical and structural properties of some elastomeric materials was investigated in the last chapter. How swelling elastomers behave when they are used as sealing elements in downhole applications is covered in this chapter. Swelling under actual well conditions is a complex and highly non-linear problem. It is important to know how the elastomer actually swells in a particular well, how much time is required to achieve sealing, and what sealing pressure is generated. As mentioned earlier, design and development of swelling elastomer seals is mostly based on hit-and-trial method and some laboratory tests. Analytical models for swelling of rubberlike materials are difficult to formulate, and restricted in actual application due to their need for simplifying assumptions [1–3]. Tests conducted on laboratory size samples of swelling elastomers cannot reproduce actual oil well conditions, and cannot cover all possible variations of testing parameters. However, these laboratory tests do provide useful information about material response of swellable elastomers in various conditions, serving as a basis for analytical and numerical modeling [4, 5]. Though time-consuming to run under all possible conditions, properly developed and robust numerical models can be used to predict near-actual performance of elastomeric seals.

Before this study, almost no published work is available that numerically investigates seal performance based on actual material properties and actual compression data at various stages of swelling. The current chapter describes the use of numerical (finite element) simulation to investigate swelling elastomer seal behavior in downhole petroleum applications [6]. Variations in sealing (contact) pressure are studied for seal length, seal thickness, compression ratio, water salinity, swelling time, and type of well completion (open-hole or cased-hole). Month-long swelling experiments on samples of two actual elastomers (Chapters 3 and 7) provide input to the numerical model in terms of real material and deformation data. On the basis of these results, petroleum engineers can make informed decisions about the selection of elastomer material and seal geometry appropriate for the well type and conditions encountered. Application developers and researchers can also find this investigation useful in performance analysis and design of swelling elastomer seals.

2. Mechanism of sealing

In oil wells, swelling elastomer seal is placed in the drilled hole with a certain gap between the seal and the outer casing (or formation), to ensure easy placement. With exposure to swelling fluid, the elastomer swells freely, the only driving force being dynamic swelling; through osmosis for water swellables [7], and through diffusion for oil swellables [8]. It first fills the gap between the tubular and formation; this may take a few hours to a few days depending on the elastomer type Further swelling (after filling up of the gap) generates pressure at the interface. Sealing is effective when contact pressure generated by swelling is greater than the differential fluid pressure across the seal ends in the well [9]. Swelling elastomers may be deployed in either open-hole or cased-hole completions. In open-hole configuration, the elastomer is in direct contact with the formation, while in cased-hole completions, the elastomer makes contact with the outer tubular or casing. It was concluded in Chapter 6 that among the existing hyperelastic material models, Ogden model with second degree strain energy function [10] most closely represents the behavior of swelling elastomers. Ogden-2 is therefore used in this chapter for numerical simulation of seal performance.

3. Material behavior: experimental work

As described in previous chapters, experiments were designed and conducted to find out material properties and coefficients needed for numerical simulation of swelling elastomer seal behavior. Disc samples of two elastomer types (designated as material-*A* and material-*B*) were allowed to swell for a 30-day test period in salt-water solutions of two different concentrations (6,000 and 120,000 ppm), maintained at a temperature of 50°C. Test temperature and salinities were selected to emulate medium-depth wells in the region. Swelling measurements included variations in volume, thickness, hardness, and density. Compression and bulk tests were performed to determine mechanical properties (elastic modulus, bulk modulus, Poisson's ratio, etc). Due to the quick-swelling nature of the elastomers, readings were initially taken after each day, and with increasing time periods later on (after 1, 2, 4, 7, 10, 16, 23, and 30 days of swelling).

To get the properties of the steel tubular on which the elastomer sections were mounted, strips were cut along the length of the tubular. Milling and grinding machines were then used to prepare tensile samples in line with ASTM E8M [11]. An Instron universal testing machine (150 kN) was used to conduct the tests. For the numerical simulation, coefficients for respective material models were extracted by curve fitting of the test results: stress–strain data from compression tests and pressure-strain data from bulk tests conducted on elastomer materials, and stress–strain data from tensile tests on tubular steel material. For full details of the experimental work, please refer to Chapter 7.

4. Numerical modeling and simulation

Modeling of the compression of an elastomer seal against an outer casing or a rock formation is treated as a static problem. Arrangement of the steel tubular, elastomer seal segment, and casing/formation in the pre-compression stage is shown schematically in **Figure 1**. An axi-symmetric model is used, and all materials are considered to be deformable. FE modeling is done on the software package ABAQUS [12]; CAX4RH element (4-node bilinear axi-symmetric quadrilateral hybrid) with reduced integration is used to represent the elastomer seal; and the steel tubular and rock formation are modeled using the CAX4R element (4-node bilinear axi-symmetric quadrilateral) is used to represent. In actual packers, elastomer seal element is vulcanized onto the tubular, therefore tubular and elastomer are treated as perfectly bonded in the model. Interaction forces at the

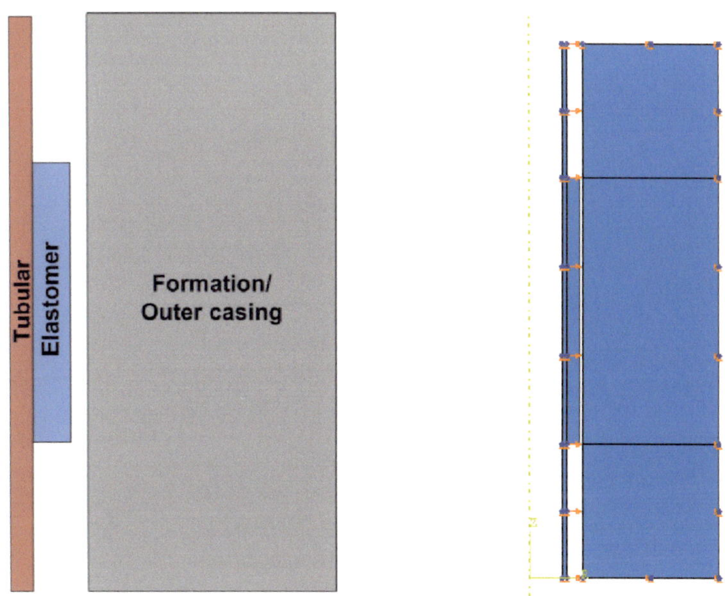

Figure 1.
Schematic diagram of model with tubular, elastomer, and formation (left); and axi-symmetric numerical model with boundary conditions (right).

elastomer-formation contact surface are split into tangential and normal components. Coulomb law is used to describe friction, with a friction coefficient of 0.4 in the tangential direction, and hard contact in the normal direction. If there is an outer casing in place of the rock formation (cased-hole), coefficient of friction is reduced to 0.1 because of the improved surface finish.

Tubular thickness and length are 7.5 mm and 800 mm respectively, elastic modulus is 210 GPa, and Poisson's ratio is 0.3. Elastomer seal is 29 mm thick and 400 mm long (same as actual swellable packer), while length variation of 100–600 mm is used to study the effect of seal length. Coefficients of Ogden hyperelastic material model are evaluated using experimental results from Chapter 6. Based on actual properties in a regional oil well, formation is modeled using Mohr Coulomb plasticity having following values: friction angle of 18°, dilatation angle of 15°, and cohesion yield strength of 7.515 GPa. For the outer steel casing, Young's modulus of 207 GPa and Poisson's ratio of 0.3 have been used.

Figure 1 also shows the ABAQUS model with all boundary conditions. Tubular is restricted to move in the vertical direction, while displacement is applied to the inner surface in the horizontal direction. When exposed to water or oil, the elastomer swells freely (in the thickness direction) until it fills the gap and comes in contact with the formation (or outer casing). Applied displacement amount is the difference between maximum free swelling in the thickness direction (from experiments) and the gap. Formation is fixed on all surfaces except the elastomer-formation contact surface.

After the gap is filled, the elastomer touches the formation and starts generating pressure. This contact pressure will develop the seal between elastomer and formation. Upper half of undeformed and deformed ABAQUS model is shown in **Figure 2**. It can be seen that after pressurization, the elastomer slides along the formation; this phenomenon occurs at the lower end also (not shown in the figure). The retaining end rings in an actual packer support the elastomer and do not allow it to slide along the gap in this way.

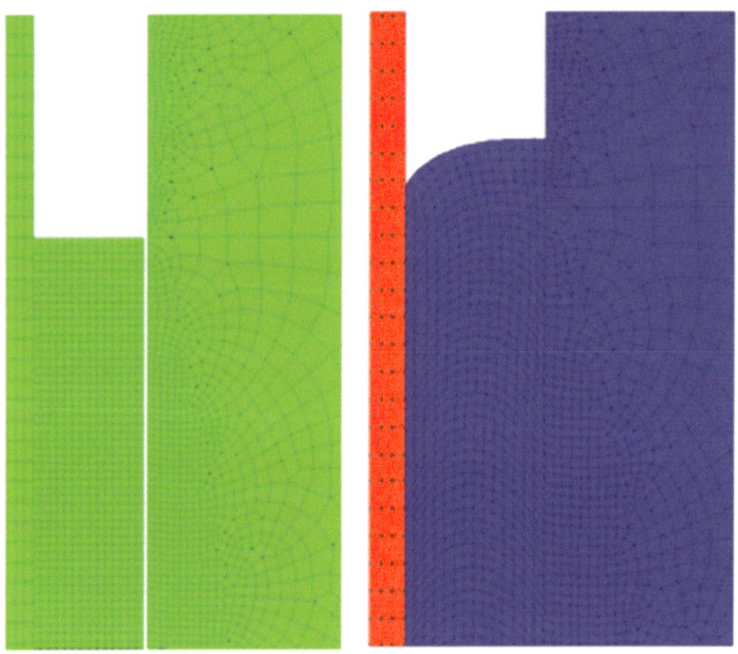

Figure 2.
Elastomer seal before (left) and after deformation (right).

5. Results and discussion

As discussed earlier, there are two factors that are necessary for superior sealing: higher value of contact pressure at the elastomer-casing interface; and good integrity and reliability of the elastomer material under the given conditions. Extended exposure to saltwater (or other medium) makes the elastomer softer (decreased hardness). Effective service life of the elastomer seal in the face of this softening effect is an important issue. Longevity investigation over a 5-year test period for a variety of water-type and oil-type swell packers has already been presented in Chapter 5.

Effect of the elastomer seal material and geometry, and the testing parameters, on seal contact pressure is presented graphically and discussed in the following sections. An expected, and notable, variation pattern can be observed in all the graphs. Seal pressure is lowest at the two ends of the rubber element, and increases nonlinearly to a maximum value near the central location.

5.1 Effect of swelling period

Figure 3 shows variation of seal contact pressure over the seal length after different swelling periods for two elastomer materials in low salinity brine. As observed for laboratory samples (Chapter 3), longer exposure results in larger amount of swelling. More swelling (increased seal thickness) generates more seal compression against outer casing or formation; increasing the contact pressure. As mentioned above, these elastomers are fast-swell type: most of the swelling occurs in the initial days, and the swelling rate goes down afterwards. Because of this, the well can become operational earlier, a highly sought-after outcome in petroleum engineering because of financial considerations. Faster swelling rate generates higher sealing pressure in the earlier days, and slower swelling for longer periods leads to lower increase in seal pressure later on.

More swelling leads to higher contact pressure, increasing the strength of the seal. **Figure 4** shows the variation in seal contact pressure for both materials in high salinity brines for the entire one month period. Progressive swelling increases the sealing contact pressure and strengthens the seal. Maximum pressure attained in each case (at the middle of the sealing element) is used for this plot. As observed and explained above, large increase in contact pressure occurs only in the first seven days; then the pressure increase becomes more gradual. A little fluctuation in the swelling pattern can be observed for material-*B* near the end of the first week. A

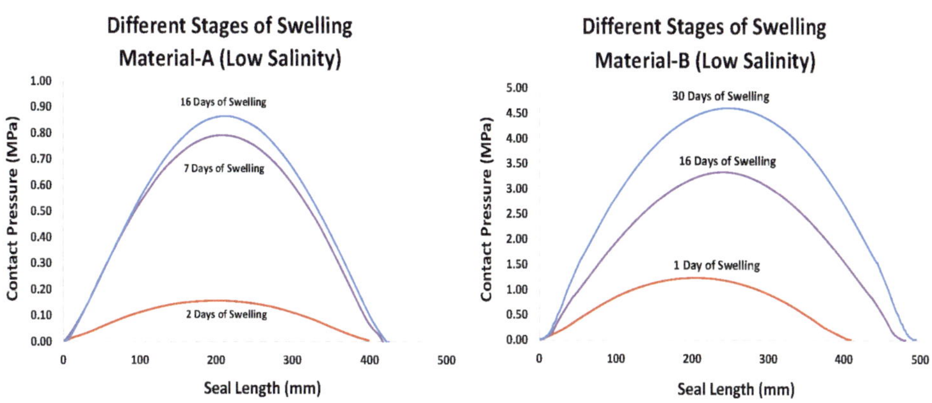

Figure 3.
Seal pressure variation for both materials at the end of different swelling periods under low salinity.

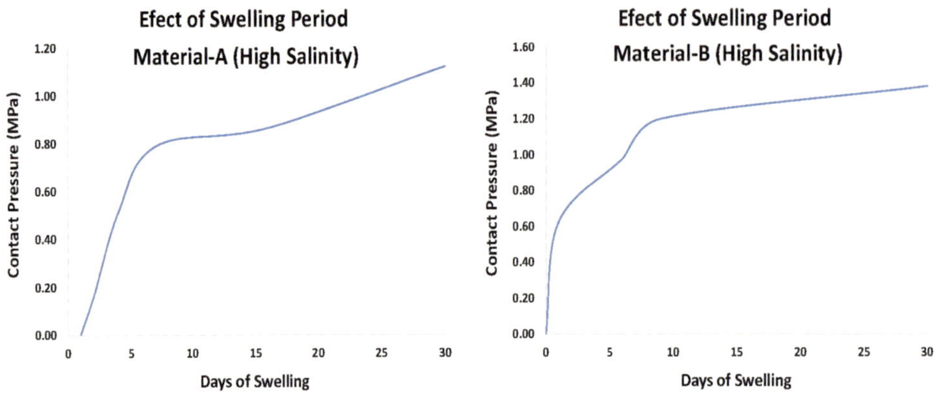

Figure 4.
Effect of swelling on sealing pressure for the entire period; both materials; high salinity.

rationale for this slight irregularity, especially for swelling under higher-salinity brine, has been presented in some earlier works by the authors [5, 13, 14]: two-way transport of salt between elastomer and salt solution, and breaking and re-forming of polymer cross-link chains. As far as choice of elastomer type goes, both materials show almost the same seal pressure performance in the case of higher-salinity wells.

5.2 Effect of water salinity

Figure 5 shows sealing pressure curves for low and high water salinity after seven days of swelling for material-A, and after thirty days of swelling for material-B. Pressure curve for lower salinity is higher for both materials, as expected. It was found in Chapters 3, 6, and 7 that larger amount of swelling (thickness increase) takes place in lower-salinity water; this will generate larger compression ratio and thus higher contact pressure. The reason is that higher chemical potential gradient is present in the case of low-salinity solution, giving rise to more fluid influx into the elastomer at a higher rate, as explained earlier.

5.3 Pressure variation pattern

Variation trend of sealing pressure against swelling rate is shown in **Figure 6**, for the two salt concentrations. The technique is to tag the center point of the elastomer,

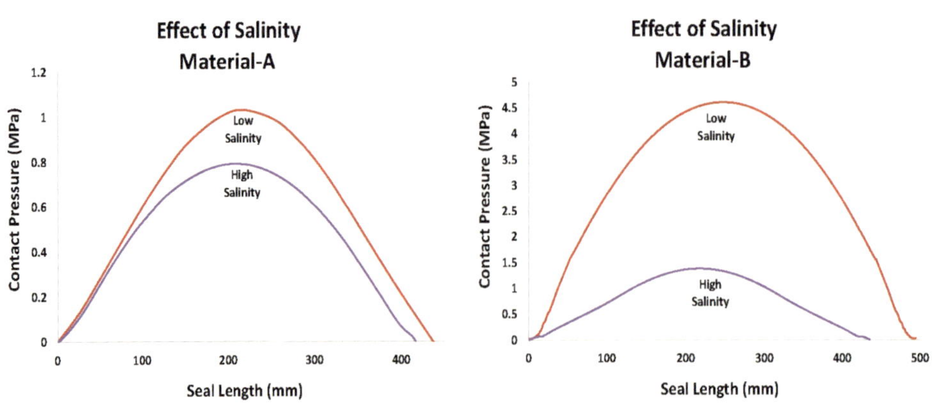

Figure 5.
Effect of salinity on seal contact pressure; material-A, 7 days of swelling; material-B, 30 days of swelling.

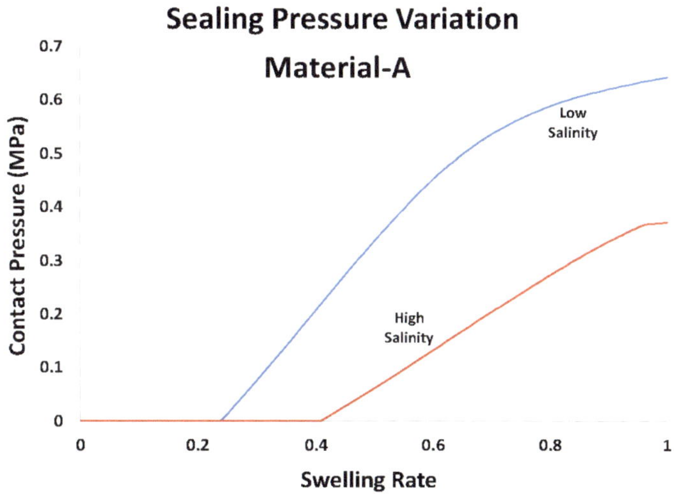

Figure 6.
Pressure variation pattern for material-A; low and high salinities.

where maximum pressure occurs, and then record the change in pressure as the amount of swelling increases. As mentioned earlier, the elastomer keeps on swelling until it just touches the formation; further swelling causes it to exert increasing pressure because of the compression. This increase in pressure is faster at first, and then drops down to a more gradual change. Elastomer swelling is higher in low salinity brine, and therefore has the higher pressure curve. For both salinities, pressure variation pattern is similar, but there are obvious differences in the magnitude of contact pressure generated.

5.4 Effect of seal length

In order to investigate the effect of length of sealing element on contact pressure, simulations are run for lengths of 100 mm, 200 mm, 300 mm, 400 mm, 500 mm and 600 mm with same set of boundary conditions. Maximum pressure obtained from each simulation is plotted in **Figure 7**; after three days of swelling for material-A, and after seven days of swelling for material-B in high salinity solution. Increase in seal length gives higher contact pressure. As expected, for a fixed amount of seal compression, larger contact area between elastomer seal and outer casing yields higher sealing pressure. As for the slight fluctuation of contact pressure for material-*B* (under high salinity), it may be due to the factors related to the nature of swelling described above.

5.5 Effect of seal thickness

How thick the elastomer seal element should be is an important decision in downhole petroleum applications. Based on actual field data, different thicknesses (10 mm to 29 mm) are used to investigate the effect of seal thickness on contact pressure. As an example, **Figure 8** shows simulation results for different seal thickness for both materials under low salinity after seven days of swelling. One observation appears to be counter-intuitive: pressure is higher for smaller seal thickness. The reason is the way the simulation is conducted: magnitude of seal compression is kept constant, while simulation is run for different values of seal thickness. Compression ratio (seal compression δ divided by seal thickness t) is therefore larger for thinner seals; thus the observed behavior. This

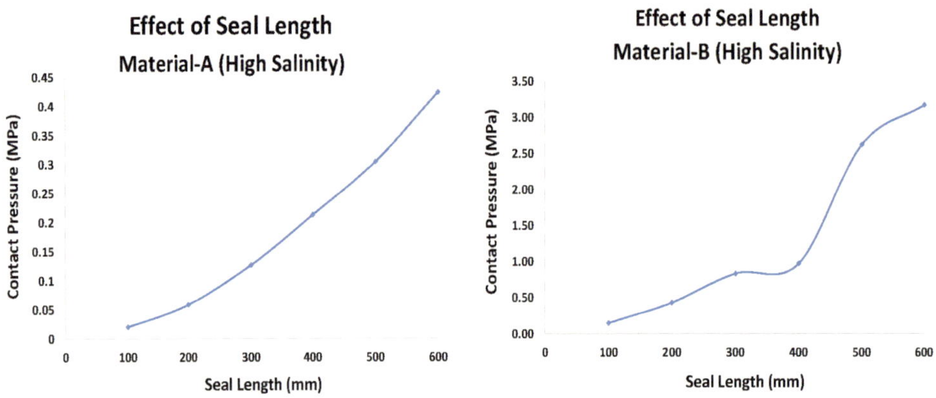

Figure 7.
Effect of seal length on seal contact pressure; high salinity; material-A, 3 days of swelling; material-B, 7 days of swelling.

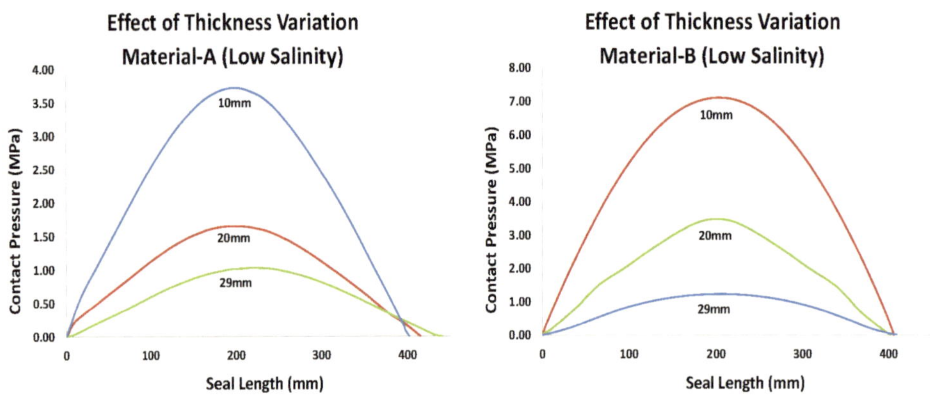

Figure 8.
Effect of seal thickness on sealing pressure; both materials; low salinity; 7 days of swelling.

misperception can be avoided by using the expression *compression ratio* instead of *seal compression*; as done in the following sub-section.

5.6 Effect of compression ratio

Figure 9 shows the variation of maximum sealing pressure versus compression ratio in for both materials in low salinity solution after seven days of swelling. As expected, sealing pressure is higher when larger compression ratio is larger. In actual downhole operations, higher value of seal pressure is naturally preferred, as it represents better seal performance. In applications where swellable elastomers and solid expandable tubulars (SET) are used conjointly, compression ratio is a result of elastomer swelling, or tubular expansion (larger expansion ratio), or a combination of the two [15, 16].

5.7 Effect of well completion type

As explained in Chapter 2, swelling elastomers are used in many applications for remediation as well as new installations using both open-hole and cased-hole completion strategies [17–19]. In selecting a suitable swell packer, it is important to consider the effect of well completion type on contact pressure. **Figure 10** shows

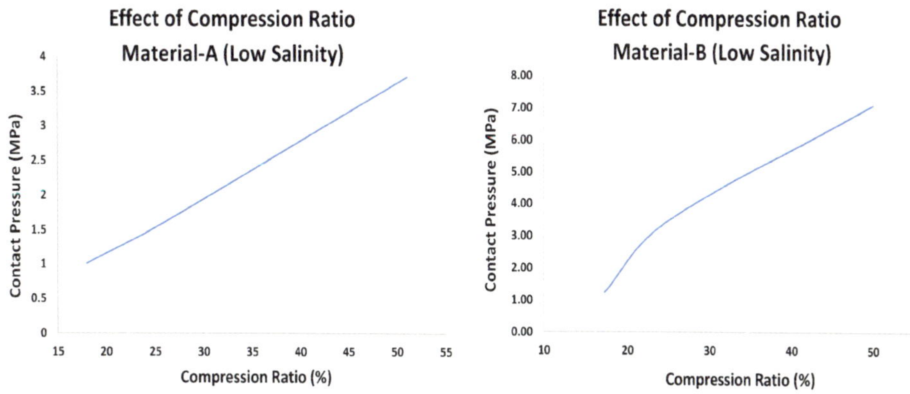

Figure 9.
Effect of compression ratio on seal contact pressure; both materials; low salinity; 7 days of swelling.

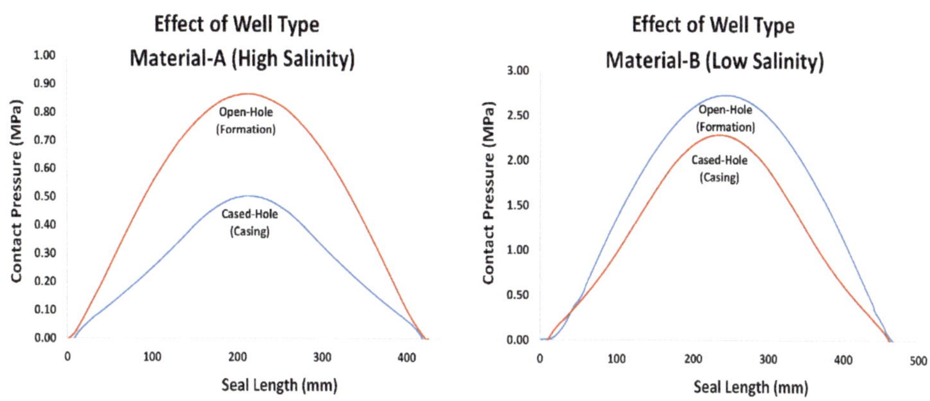

Figure 10.
Effect of well completion type on seal contact pressure; material-A, high salinity, 16 days of swelling; material-B, low salinity, 9 days of swelling.

this in a graphical format for material-*A* after 16 days of swelling in saltwater of higher concentration, and for material-*B* in low concentration solution after 9 days of swelling. These simulation results are in good agreement with actual behavior in oil wells. In open-hole configuration, when elastomer is in direct contact with rock formation, there is higher contact friction due to larger surface roughness of the formation; sealing pressure is therefore higher. For cased-hole completions, the elastomer seal makes contact with the outer casing, which is smooth and has much better surface finish compared to the formation. The lower friction coefficient yields lower sealing contact pressure.

5.8 Selection of appropriate seal

Three factors are critical in proper selection of swell packer or elastomer type in actual oil or gas wells: rate of swelling, as it influences how early full sealing is achieved and the well goes into production; magnitude of seal contact pressure, as it is needed to withstand the differential pressure in the well; and durability (long-term integrity) of the elastomer seal. On the basis of the various simulation results described above, some significant observations can be made about the proper choice of elastomer seals for actual oil well conditions: (a) If the well is a higher-salinity type, magnitude of swelling would be small, and ensuing seal pressure would be low. This seal will not be effective against large differential pressures. Furthermore, rate of swelling will be low, requiring longer well completion time, leading to time and cost overruns. (b) It appears that larger seal length yields higher sealing pressure. However, very large seal length is not practical: longer packers will be more costly, and their deployment in the wells would be more difficult. A much better design would be to have a series of well-spaced elastomer sections on the steel tubular rather than a single-element seal of very large length. (c) Seals of larger thickness will naturally generate larger pressures as they will fill the small annular gap quickly. Nevertheless, deploying a large-thickness packer to the target location in the well can cause obstruction related problems. A minimum amount of annular clearance has to be kept to insert measurement and remediation tools when needed. If quick water shutoff is required, fast swelling elastomer under the given temperature and salinity conditions should be used with an optimally large seal thickness. Optimum packer design needs a judicious balance of elastomer type (fast-swelling) and a seal thickness that is not too large. (d) In the case of an open-hole completion, roughness of the rock formation helps generate higher contact pressure; elastomers that are not fast-swelling may be used to good effect.

6. Conclusions

Finite element simulations have been carried out to investigate sealing performance of downhole applications, for two different swelling elastomer materials that are kept in low and high salinity brine solutions. Required experimental data about swelling and compression tests is reported in Chapters 3 and 7. Finite element simulations are conducted to study the performance of elastomer seals (contact pressure) under different parameters such as well completion type (open-hole or cased-hole), seal configuration (length, thickness), downhole conditions (saltwater concentration, temperature, differential pressure), swelling time, and compression ratio. Contact pressure increases with the amount of swelling (or swelling time), at a faster rate in the first few days, and more gradually later on. Pressure curves are higher for low-concentration brine. Seals of larger length yield higher contact pressure. Larger values of compression ratio (not necessarily thicker seals) generate higher sealing pressure. Higher pressures are observed in the case of the elastomer seal swelling against rock formation (open hole) rather than an outer steel casing (closed hole).

This investigation can benefit petroleum engineers in proper selection of elastomer type and seal geometry for existing conditions in an oilfield. Reported results can also help in performance assessment and design improvement of swelling elastomer seals and packers.

Author details

Sayyad Zahid Qamar[1]*, Maaz Akhtar[2] and Tasneem Pervez[1]

1 Mechanical and Industrial Engineering Department, Sultan Qaboos University, Muscat, Oman

2 Mechanical Engineering Department, N.E.D. University of Engineering and Technology, Karachi, Pakistan

*Address all correspondence to: sayyad@squ.edu.om

References

[1] Al-Hiddabi SA, Pervez T, Qamar SZ, Al-Jahwari FK, Marketz F, Al-Houqani S, van de Velden M (2015) "Analytical Model of Elastomer Seal Performance in Oil Wells," Applied Mathematical Modelling, 39 (10-11), June 2015, p 2836-2848

[2] Al-Kharusi MSM, Qamar SZ, Pervez T, Akhtar M (2011) "Non-Linear Model for Evaluation of Elastomer Seals Subjected to Differential Pressure," Paper # SPE-149032, SPE/DGS Saudi Arabia Section Technical Symposium and Exhibition, 15-18 May 2011, Al-Khobar, Saudi Arabia

[3] Al-Kharusi MSM, Qamar SZ, Pervez T, Akhtar M (2013) "Elastomer Seal with Frictional Contact — Analytical Solution," ASME-2013 International Mechanical Engineering Congress and Exhibition (IMECE-2013), 15-21 November 2013, San Diego, California

[4] Qamar SZ, Al-Hiddabi S, Pervez T, Marketz F (2009) "Mechanical Testing and Characterization of a Swelling Elastomer," Journal of Elastomers and Plastics, 41 (5), p 415-431

[5] Qamar SZ, Pervez T, Akhtar M (2016) "Performance Evaluation of Water-Swelling and Oil-Swelling Elastomers," Journal of Elastomers and Plastics, 48 (6) 2016, p 535-545

[6] Akhtar M, Qamar SZ, Pervez T, Al-Jahwari FK (2018) "Performance Analysis of Swelling Elastomer Seals," Petroleum Science and Engineering, 165 (2018), p 127-135

[7] Qamar SZ, Akhtar M, Pervez T (2015) "Swelling Behavior of Soft and Biological Materials," Key Engineering Materials, Vol 656-657 (2015), p 580-585

[8] Akhtar M, Qamar SZ, Mehdi SM, Hussain A (2018) "Diffusion-Based Swelling in Elastomers under Low and High-Salinity Brine," Journal of Elastomers and Plastics, 51 (1), p 75-84,

[9] Akhtar M, Qamar SZ, Pervez T, Al-Jahwari FK (2013) "FEM Simulation of Swelling Elastomer Seals in Downhole Applications," ASME-2013 International Mechanical Engineering Congress and Exhibition (IMECE-2013), 15-21 November 2013, San Diego, California

[10] Ogden R (1972) "Large Deformation Isotropic Elasticity - On the Correlation of Theory and Experiment for Incompressible Rubberlike Solids," Proceedings of the Royal Society of London A. Mathematical and Physical Sciences, 326 (1567), p 565-584

[11] ASTM E8/E8M-15a (2015) "Standard Test Methods for Tension Testing of Metallic Materials," ASTM International, West Conshohocken, PA

[12] ABAQUS (2012), Software version 6.11, Dassault Systems

[13] Qamar SZ, Akhtar M, Pervez T, Al-Kharusi MSM (2013) "Mechanical and Structural Behavior of a Swelling Elastomer under Compressive Loading," Materials and Design, 45 (March 2013), p 487-496

[14] Qamar SZ, Pervez T, Akhtar M, Al-Kharusi MSM (2012) "Design and Manufacture of Swell Packers: Influence of Material Behavior," Materials and Manufacturing Processes, 27 (7), 2012, p 727-732

[15] Alzebdeh K, Pervez T, Qamar SZ (2010) "Finite Element Simulation of Compression of Elastomeric Seals in Openhole Liners," Journal of Energy Resources Technology, 132 (3), Sept 2010

[16] Velden M, Al-Jahwari F (2013) "Expansion of Tubular with Elastomers

in Multilateral Wells," The Journal of Engineering Research, 10 (1), p 41-49

[17] Al-Balushi HM, Al-Rashdi MH, Al-Shandoodi MH, van Noort RH (2004) "World's Longest Expandable Open Hole Clad and Open Hole Liner with Swelling Elastomer Deployed in Yibal Horizontal Well", Proceedings IADC/SPE Asia Pacific Drilling Technology Conference and Exhibition, SPE Paper # 88026-MS

[18] Al-Douseri KM, Barnes CG, Young DA, Smith PE (2009) "Swellable Packers Provide a Brownfield Water Management Solution in Open and Cased Hole," Paper # SPE-124394, Offshore Europe, 8-11 September, Aberdeen, UK

[19] Ekpe J, Kompantsev A, Al-Thuwaini J, Belousov Y, Ashoor A, Abdul-Aziz ME (2011) "First Multistage Swell Packers Openhole Completion in Rub al-Khali Empty Desert: Case Study," Paper # SPE 142523, SPE Middle East Unconventional Gas Conference and Exhibition, Muscat, Oman,

Chapter 9

Swelling Elastomers and Tubular Expansion—Numerical Investigation

Sayyad Zahid Qamar, Maaz Akhtar and Tasneem Pervez

Although this may seem a paradox, all exact science is dominated by the idea of approximation. When a man tells you that he knows the exact truth about anything, you are safe in inferring that he is an inexact man.

Bertrand Russell

Abstract

Swell packers were initially used for repair of old and damaged wells, but they are now increasingly used for higher productivity and profitability through developments like slim well design and reduced-cement or cementless completions. Solid expandable tubular (SET) technology has gained popularity in the petroleum development industry as it can reduce well costs and improve well performance. A conical mandrel is pushed or pulled through a petroleum tubular, either hydraulically or mechanically, to expand it (in-situ) to the desired diameter. In SET applications such as water shutoff and zonal isolation, swelling elastomers are an obvious choice as a sealing material. For proper downhole deployment of swell packers in SET applications, it is important to have a good idea about their behavior under a given set of field conditions. Design and manufacturing of SET applications using swelling elastomers as sealing elements also needs some sort of seal performance analysis.

Keywords: solid tubular expansion, swell packers, seal performance, finite element analysis

1. Introduction

Swell packers were initially used for repair of old and damaged wells, but they are now increasingly used for higher productivity and profitability through developments like slim well design [1] and reduced-cement or cementless completions [2]. Solid expandable tubular (SET) technology has gained popularity in the petroleum development industry as it can reduce well costs and improve well performance [3, 4]. A conical mandrel is pushed or pulled through a petroleum tubular, either hydraulically or mechanically, to expand it (in-situ) to the desired diameter [5–7]. In SET applications such as water shutoff and zonal isolation, swelling elastomers are an obvious choice as a sealing material [8, 9]. For proper downhole deployment of swell packers in SET applications, it is important to have a good idea about their behavior under a given set of field conditions [10]. Design and manufacturing of SET applications using swelling elastomers as sealing elements also needs some sort of seal performance analysis.

Schematic layout of tubular expansion against rock formation or outer steel casing, together with swelling elastomer seals, is shown in **Figure 1**. Compression of elastomer against the formation provides the required sealing [11]. This compression can be achieved through expansion of the inner tubular, by swelling of the elastomer element against the formation [12], or through a combination of the two. In the usual case, when one end of the SET is anchored and the other is free, the expansion process is known as fixed-free. In the rare occurrence when both ends are fixed, due to some aberrations in the expansion process (such as sticking of the tubular to the formation at some point), the condition is termed as fixed-fixed.

This chapter reports results of numerical (finite element) simulation used to investigate the effect of various field conditions on the sealing or contact pressure generated between the elastomer element and the rock formation [13]. This is the first study of its kind, investigating seal behavior based on material properties of expandable tubular and different swelling elastomers used in actual oil wells. These parameters include elastomer seal material, seal thickness, SET expansion ratio, elastomer compression ratio, formation type (rigid, elastic, elastic-plastic), and SET boundary conditions (fixed-free or fixed-fixed). Results of this work could be used by field engineers, application developers, and researchers for performance analysis and design improvement of solid expandable tubulars and swelling elastomer applications, and for proper selection of packers for a given set of field conditions.

1.1 Elastomer seal material

Various swelling elastomers have been developed to suit different types of well conditions (water salinity, temperature, differential pressure, etc). Choice of elastomeric material will definitely have an impact on seal performance. Based on experimental studies conducted by the authors on material characterization of

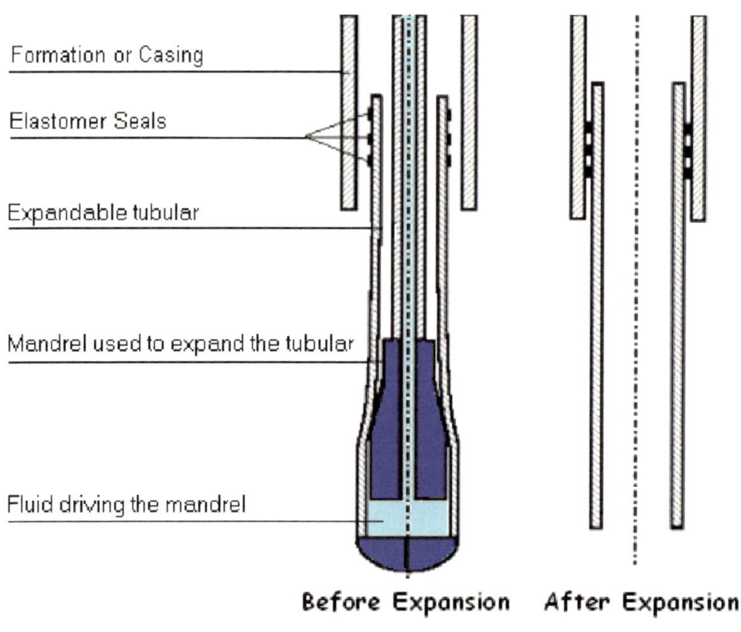

Figure 1.
Schematic illustration of tubular expansion with swelling elastomer seals.

swelling elastomers used in the regional oil fields [14, 15], five elastomer materials are used in this study. These seal materials are referred to as elastomer-1 (E1), elastomer-2 (E2), etc. Average values from three samples of each elastomer are used as input for the numerical model. A typical three-sample stress-strain curve for one of the elastomers is shown in **Figure 2**. Averaging out the three curves for each elastomer, **Figure 3** shows the representative stress-strain behavior for the five elastomer materials used in the study.

1.2 Expansion ratio (ER) and compression ratio (CR)

By using mandrels of different size, different expansion ratios can be achieved for a tubular of the same initial diameter [16]. Expansion ratio (ER) describes the percentage increase in tube diameter due to the expansion process

$$ER = \frac{OD_m - ID_t}{ID_t} * 100. \quad (1)$$

OD_m is the outer diameter of the mandrel (cone), and ID_t is the inner diameter of the pre-expanded tubular (SET). More important in terms of the seal contact pressure generated is the elastomer compression ratio (CR) that describes the amount of compression of the seal. This is the combined effect of expansion of the steel tubular, and swelling of the elastomer seal when exposed to a swelling medium.

$$CR = \frac{ER * R_t - G}{t} * 100. \quad (2)$$

This can also be expressed in terms of the amount of expansion of the tubular as

$$CR = \frac{(OR_m - IR_t) - G}{t} * 100. \quad (3)$$

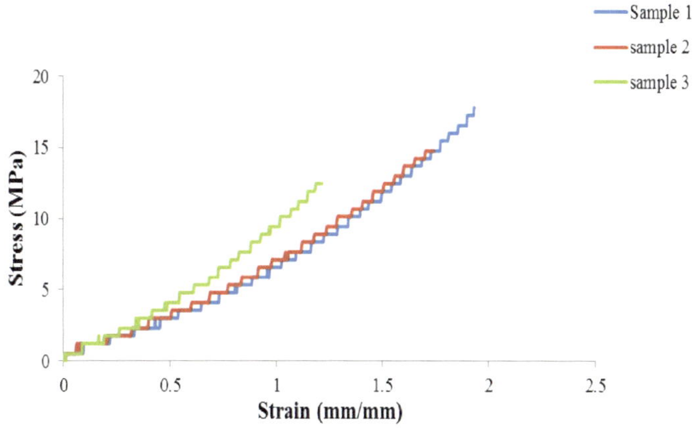

Figure 2.
Three-sample stress-strain graph for elastomer-3.

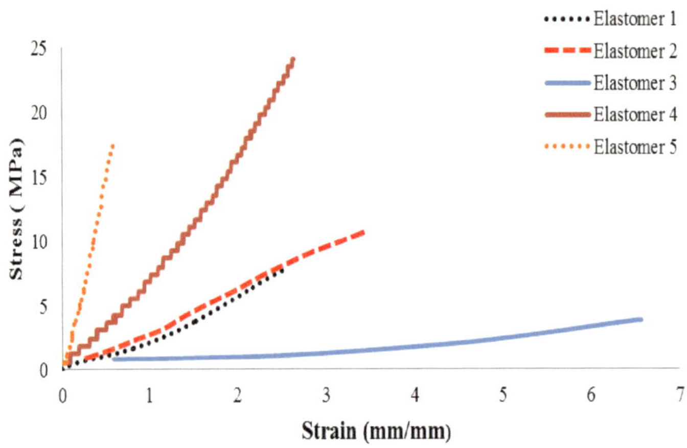

Figure 3.
Average stress-strain curves for the five elastomeric seal materials (right).

OR_m is the outer radius of the mandrel, IR_t is the inner radius of the SET before expansion, G is the initial radial gap between the elastomer seal and formation, and t is the seal thickness.

2. Numerical modeling

Compression of the SET elastomer seal against rock formation is modeled and simulated using the commercial finite element package ABAQUS; **Figure 4**. Based on the geometric layout of the problem, an axisymmetric model is used. The mandrel is moved through the tubular, expanding it and thus compressing the elastomer seal against the formation. Arruda-Boyce hyperelastic model is chosen to represent the material behavior of the elastomer. Elastomer is assumed to be perfectly bonded to the tubular. Coulomb friction model is used to represent the mandrel-tubular and seal-formation surface interactions. Formation is modeled as rigid, elastic, or elastic plastic. Seal length is varied from 50 mm to 400 mm, as seal lengths below 50 mm are not economically viable for either the manufacturers or the operators. Based on actual swell packer data, elastomer seal thickness of 5, 7, or 9 mm is modeled.

The expansion cone (mandrel) is made of hardened, heat-treated tool steel. It is therefore modeled as a rigid body. The tubular is modeled using 4-node bilinear axisymmetric quadrilateral elements with reduced integration. Dimensions of an actual 7⅝-inch solid expandable tubular are used; inner diameter of 174.625 mm and thickness of 9.525 mm. Material properties are obtained from tensile test data for actual unexpanded SET samples. Compared to normal steels, this expandable steel has an unusual elastic modulus of around 78 GPa. Poisson's ratio is taken to be 0.3.

Element type used for the elastomer seal is 4-node bilinear axisymmetric quadrilateral with hybrid formulation. Material coefficients for the Arruda-Boyce hyperelastic model for each elastomer are extracted from stress-strain data obtained from tensile tests conducted on the unswelled elastomers; **Figure 3**. As described above, the rock formation is modeled as rigid, elastic, and elastic-plastic. Formation properties are obtained from an actual oil field in the region. Variable mesh refinement is used in areas of large deformation for the tubular and elastomer seal; **Figure 4**.

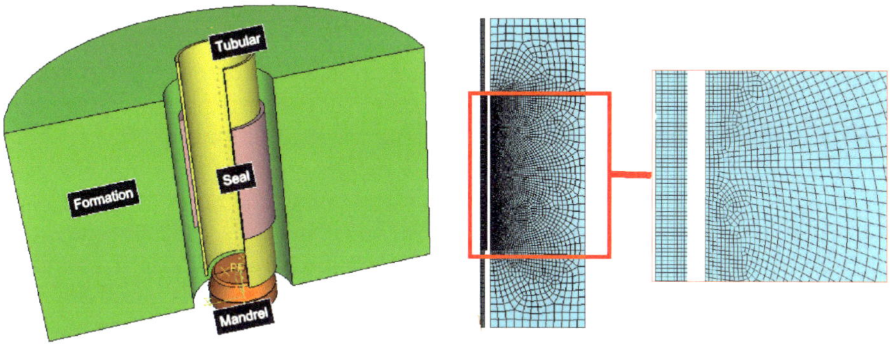

Figure 4.
Finite element model of SET expansion with swelling elastomer seal (left); meshing of the model, with enlarged view of central contact area between elastomer and formation (right).

3. Results and discussion

FE simulations have been run to study variations in seal contact pressure for different elastomer materials, expansion ratios, compression ratios, seal thickness, formation type, and tubular end conditions.

3.1 Elastomer material

Figure 5 shows the effect of using different swelling elastomers on sealing pressure along the seal length. Except near the ends of the sealing element, there is no significant pressure variation along the length of the seal. The ends have more freedom to deform, while the rest of the elastomer (along the entire seal length) is perfectly bonded to the steel tubular, so this observation is as expected. Elastomer E5 generates the highest contact pressure, while E3 yields the lowest. This result is in line with the fact that elastomer-5 has the highest modulus of elasticity, and the stiffer material produces larger forces. It is important to note that these results are based on elastomer properties in the un-swelled condition. When the elastomers swell, sealing pressure will be higher due to the combined effect of tubular expansion and elastomer swelling. Selection of the elastomer type for a particular SET application will depend on the actual field conditions such as temperature, water salinity, formation type, etc.

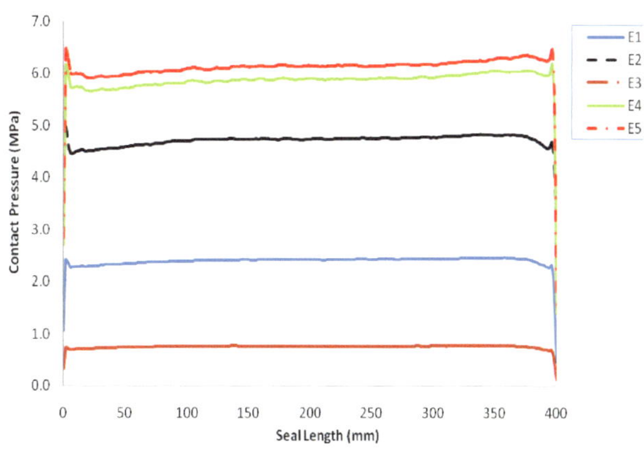

Figure 5.
Contact/sealing pressure along the seal length for different elastomer materials.

3.2 Expansion ratio

Effect of tubular expansion ratio on contact pressure along the length of the seal is shown in **Figure 6**. Obviously, larger amount of tubular expansion (higher value of *ER*) pushes the elastomer more against the formation, generating higher sealing pressure. Behavior along the seal length is interesting; contact pressure at the two ends is much higher, then drops down sharply and remains almost constant along the middle portion of the seal. The mandrel is conical in shape. Pressure starts to build up as the mandrel comes in contact with the tubular, and builds up sharply as the rest of the conical surface pushes against the tubular. After this initial build-up, the pressure drops down to almost a steady-state value. Mostly dependent on the variation in frictional forces (sticking to sliding), this initial rise and drop in pressure (known as *upsetting*) is typical of the process of extrusion, which is very similar to tubular expansion. Near the end of the seal, when the conical mandrel starts to lose contact with the tubular, a mirror behavior can be clearly observed.

3.3 Seal thickness

Effect of seal thickness on sealing contact pressure can be seen in **Figure 7**. Pressure curve for the thickest seal (9 mm) is the highest. Obviously, as the thickness increases, the compression ratio increases, generating higher contact pressure. In reality, this does not translate into the simplistic rule of *larger seal thickness for higher seal pressure*. Larger thickness of sealing elements will result in larger consumption of swelling elastomer, adding to both cost and weight. More thickness can also cause clearance issues for in-situ measurement devices. Moreover, too much sealing pressure can cause undesirable distortions. Optimum seal thickness will depend on the specific field conditions, elastomer material being used, and the amount of tubular expansion.

3.4 Formation type

Variation in sealing pressure along the seal length for different types of formation is shown in **Figure 8**. Highest contact pressure can be observed in the case when the formation is assumed to behave as a rigid material, and the lowest for elastic-plastic behavior. These observations are in line with intuitive reasoning. In the case of elastic and elastic-plastic formations, some amount of energy will be spent in elastic or plastic deformation of the formation. However, all energy will be

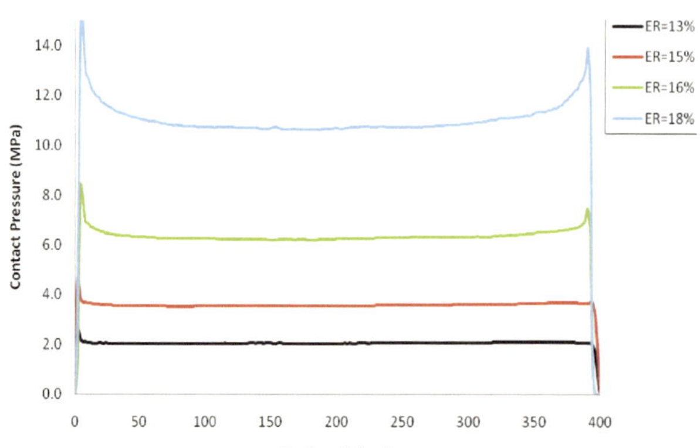

Figure 6.
Contact pressure along the seal length for different tubular expansion ratios.

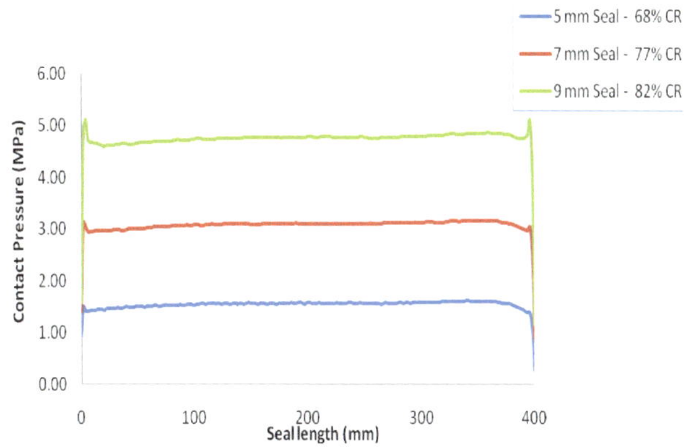

Figure 7.
Contact pressure along the seal length for different values of seal thickness.

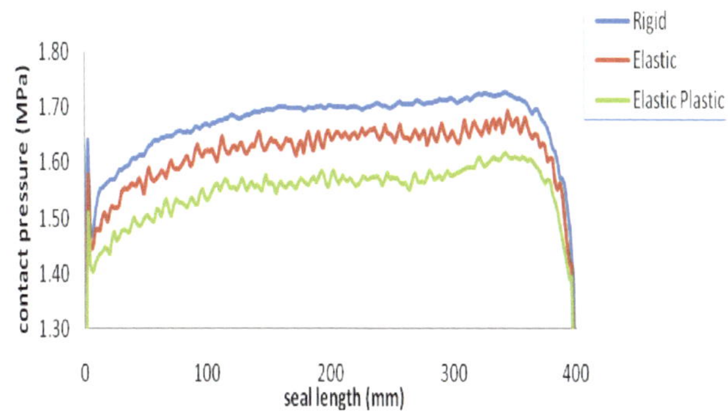

Figure 8.
Contact pressure along the seal length for different formation types.

used up for elastomer compression when we have a rigid formation, thus generating maximum contact pressure.

4. Conclusions

Behavior of swelling elastomer seals, when used in conjunction with solid expandable tubulars, has been simulated using the finite element package ABAQUS. Seal performance (in terms of contact pressure generated) was studied for variations in parameters such as elastomer material, tubular expansion ratio, seal thickness, tubular end conditions, and formation type. It was found that seal material can be a major factor contributing to sealing pressure. In addition, high contact pressures are observed for higher expansion ratios, larger seal thickness, and rigid formation type. This type of work on evaluation of seal performance, using material response of solid expandable tubulars and swelling elastomers actually being used in petroleum development operations, cannot be found in earlier published literature. Results from this study are useful both for proper selection and for design improvement of swell packers in SET applications.

Author details

Sayyad Zahid Qamar[1]*, Maaz Akhtar[2] and Tasneem Pervez[1]

1 Mechanical and Industrial Engineering Department, Sultan Qaboos University, Muscat, Oman

2 Mechanical Engineering Department, N.E.D. University of Engineering and Technology, Karachi, Pakistan

*Address all correspondence to: sayyad@squ.edu.om

References

[1] Panayirci M, Houette O, Brands S, Paraschiv M, French S (2019) "Slim Well Casing Design for a Deepwater Application Using a Fast and Flexible Finite Element Engine," SPE/IADC International Drilling Conference and Exhibition, 5-7 March 2019, the Hague, the Netherlands

[2] Cedillo G, Zett A, Han X, Elghonimy R, Raeesi B, Itter D, Landi N (2018) "Swellable Packer Evaluation Using Multi-Detector Pulsed Neutron Logging and Borax," SPE/ICoTA Coiled Tubing and Well Intervention Conference and Exhibition, 27-28 March 2018, the Woodlands, Texas, USA

[3] Al-Abri OS, Pervez T, Qamar SZ, Khan R (2016) "On the Performance Analysis of AHSS with an Application to SET Technology: FEM Simulations and Experimental Measurements," Thin Walled Structures, 101 (2016), p 58-74

[4] Jay N, Sarah F, Kristaq M (2010) "Power of Design: Solid Expandable Installation Sets Multiple New Records in Deep Shelf HP/HT Well," Paper # SPE/IADC 128366, IADC/SPE Drilling Conference and Exhibition, New Orleans, Louisiana, 2-4 February, 2010

[5] Pervez T, Qamar SZ, Al-Hiddabi SA, Al-Jahwari FK, Marketz F, Al-Houqani S, van de Velden M (2011) "Tubular Expansion in Irregularly Shaped Boreholes – Computer Simulation and Field Measurements," Petroleum Science and Technology, 29 (7), January 2011, 735-744

[6] Pervez T, Qamar SZ (2011) "Finite Element Analysis of Tubular Ovality in Oil Well," Advanced Materials Research, 264-265, 2011, p 1654-1659

[7] Pervez T, Seibi AC, Al-Hiddabi SA, Al-Jahwari FK, Qamar SZ, Marketz F (2007) "Solid Tubular Expansion in Horizontal Wells," Paper # 105704, 15th SPE Middle East Oil & Gas Show and Conference, Bahrain, 11-14 March, 2007

[8] Al-Saiid KA, Abdalla E, Cherif M, Elliot SP (2007) "Practical Uses of Swellable Packer Technology to Reduce Water Cut: Case Studies from the Middle East and Other Areas," Paper # 108613-MS, Offshore Europe, Aberdeen, Scotland, 4-7 September, 2007

[9] Glenn DD (2009) "Integrating Solid Expandables, Swellables, and Hydra Jet Perforating for Optimized Multizone Fractured Wellbores," Paper # 125345-MS, SPE Tight Gas Completions Conference, San Antonio, Texas, 15-17 June, 2009

[10] Alzebdeh K, Pervez T, Qamar SZ (2011) "Finite Element Simulation of Compression of Elastomeric Seals in Open Hole Liners," Journal of Energy Resources Technology, 132 (3), 2011, 1-8

[11] Qamar SZ, Pervez T (2014a) "Pre and Post-Expansion Mechanical Behavior of Expandable Tubulars," Materials Science & Technology 2014 (MS&T-2014), 12-16 Oct 2014, Pittsburgh, USA; (Vol. 2, p 1439-1444). Association for Iron and Steel Technology, AISTECH.

[12] Akhtar M, Qamar SZ, Pervez T, Al-Jahwari FK (2018) "Performance Analysis of Swelling Elastomer Seals," Petroleum Science and Engineering, 165 (2018), p 127-135

[13] Qamar SZ, Pervez T (2014b) "Numerical Simulation of Swelling Elastomer Seals in Tubular Expansion Applications," Materials Science & Technology 2014 (MS&T-2014), 12-16 Oct 2014, Pittsburgh, USA; (Vol. 3, pp. 2051-2057). Association for Iron and Steel Technology, AISTECH.

[14] Akhtar M, Qamar SZ, Pervez T, Khan R, Al-Kharusi MSM (2012) "Elastomer Seals in Cold Expansion of Petroleum Tubulars: Comparison of Material Models," Materials and Manufacturing Processes, 27 (7), 2012, 715-720

[15] Qamar SZ, Al-Haddabi SA, Pervez T, Marketz F (2009) "Mechanical Testing and Characterization of a Swelling Elastomer," Journal of Elastomers and Plastics, 41 (5), 2009, 415-431

[16] Pervez T, Qamar SZ, Seibi AC, Al-Jahwari FK (2008) "Use of SET in Cased and Open Holes: Comparison between Aluminum and Steel," Materials and Design, 29 (2008), p 811-817

Chapter 10

Analytical Model for Seal Contact Pressure

Sayyad Zahid Qamar, Maaz Akhtar and Tasneem Pervez

As far as the laws of mathematics refer to reality, they are not certain;
and as far as they are certain, they do not refer to reality.

Albert Einstein

Abstract

Swellable elastomers are used for zonal isolation and as an alternate to cementing is a new approach, resulting in significant reduction in time, cost, and weight. Very large strains, flexibility, resilience, and durability are their special features. Performance analysis is important design improvement and appropriate selection of swell packers. Experimental evaluation of swelling-elastomer seal performance can be very costly, and is not even possible in many cases. Numerical simulations (Chapters 8 and 9) can be more convenient, but computational effort and cost can be high. Development of closed-form (analytical) solutions is presented in this chapter to estimate the variation of contact pressure along the length of the elastomer seal. Major relevant parameters are properties of the material elastomer, seal configuration and size, magnitude of seal compression, and differential pressure across the seal. Numerical (finite element) modeling and simulation is also performed. There was good conformity between analytical and simulation results, validating the soundness of the analytical solution, and providing assurance that it can reliably predict the sealing response of the elastomer. A comprehensive parametric study is then conducted to assess seal performance while varying different key factors. Properties of the elastomer material (as it swells with exposure time) are required to run the analytical and the FE models. A large set of experiments were therefore designed and conducted to evaluate mechanical properties (E, G, K, and v) of the elastomer with gradual swelling (Chapters 3 and 7).

Keywords: Swelling elastomers, seal contact pressure, contact friction, analytical model

1. Introduction

In petroleum drilling and development, cementing is an established method used to hydraulically seal the steel casing from the rock formation (open hole), or the inside casing from the outside casing (closed hole). It serves the function of isolating certain zones and of preventing flow in the annular region [1]. Design and implementation of a cement job is a complex process, and many factors must be considered for its successful completion. Using swellable elastomers as a full or partial alternate to cementing is a new approach, resulting in significant reduction in time, cost, and weight [2]. Elastomers can withstand very large strains without any permanent deformation [3]. Flexibility, extensibility, resilience, and durability are the special additional features that elastomers offer [4, 5]. Without some sort of performance analysis, one cannot proceed with appropriate selection of a swell packer for a given set of field conditions, improvement of sealing design, assessment of seal integrity, etc. [6–8].

Main motivation for this work is the need for performance evaluation of swelling (and inert) elastomer seals used in petroleum applications. Experimental evaluation of swelling-elastomer seal performance can be very costly, and is not even possible in many cases [9]. Numerical simulations (Chapters 8 and 9), if validated, can be more convenient, but computational effort and cost can be high as simulations have to be run for each set of conditions [10, 11]. A well-founded analytical approach not only gives an elegant closed-form solution, but can give reasonably accurate and much faster prediction of elastomer performance under various actual oil and gas field conditions [12].

Development of closed-form (analytical) solutions is presented in this chapter to estimate the variation of contact pressure along the length of the elastomer seal. Major relevant parameters are properties of the material elastomer, seal configuration and size, magnitude of seal compression, and differential pressure across the seal [13]. Numerical (finite element) modeling and simulation is also performed. There was good conformity between analytical and simulation results, validating the soundness of the analytical solution, and providing assurance that it can reliably predict the sealing response of the elastomer. A comprehensive parametric study is then conducted to assess seal performance while varying different key factors. Properties of the elastomer material (as it swells with exposure time) are required to run the analytical and the FE models. A large set of experiments were therefore designed and conducted to evaluate mechanical properties (E, G, K, and υ) of the elastomer with gradual swelling (Chapters 3 and 7).

2. Rubber block under compressive loading

Various researchers have studied how an elastic rubber block behaves when subjected to transverse loading. Typically, a rubber block is in contact with two rigid plates, one above it, and one below it; **Figure 1**. Contact between the plates and the rubber block is full bonding or friction type [14, 15]. There will be a symmetric bulge in the central region of the rubber block if there is perfect bonding on both sides, or if the friction coefficient (sliding) on both contacting surfaces is the same. Bulging will be asymmetric otherwise. If friction is higher, amount of sliding will of course be lower. In the ideal case of frictionless contact (zero coefficient of friction), sliding will be maximum. In the extreme opposite case of bonded contact (infinite coefficient of friction), there will be no sliding. Published studies consistently show that the behavior is significantly different for rigid plates bonded above and below to an elastic layer, as compared with unbonded elastic layer [16, 17].

Majority of previous reports are based on the assumption that the strains are very small, so that problem can be modeled as a linear one. Some studies allow a little variation in friction at the contact surface, the treatment being somewhat closer to the actual case of elastomer seals in swellable packers. However, all these models have the symmetrical condition of full bonding or frictional contact on both sides. The few studies that address the more realistic case of bonding on one side and frictional contact on the other, treat the rubber material as incompressible, adopt an energy approach, and use a linear elastic model based on only the elastic modulus (E) of the seal material.

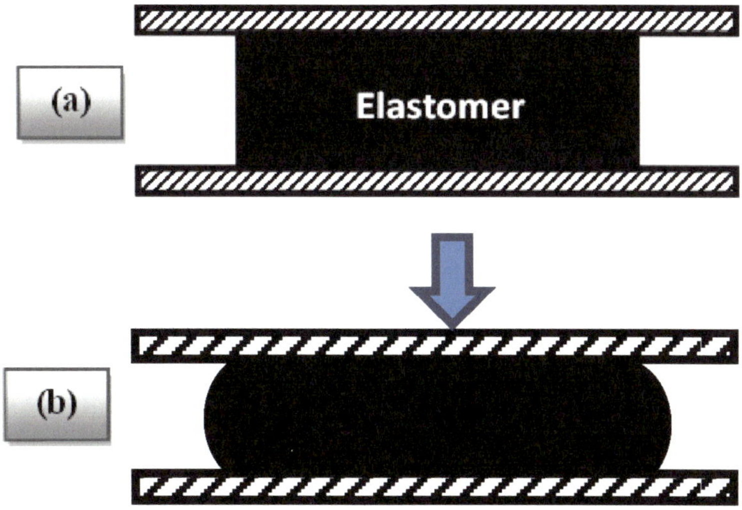

Figure 1.
Schematic diagram of deformation of elastic rubber block under transverse compressive loading; (a) before loading; (b) after loading.

2.1 Current work

None of the published studies faithfully represent the real case of inert or swelling elastomer packers used in petroleum engineering. The representative salient features that differentiate the current approach (this chapter) from all of the earlier studies are listed below.

a. Elastomer material is considered as compressible, a major shift in comparison with all earlier studies based on incompressibility (single value of Poisson's ratio: 0.5).

b. Rather than the simple geometry of flat plates, elastomer seal has a cylindrical contact on both sides, as in the real downhole situation of elastomer mounted on a steel tubular and sealing against an outer steel casing or rock formation.

c. On one side, the elastomer is fully bonded (inner steel pipe), while on the other side it has frictional contact (with outer steel casing or formation).

d. A volume-change approach is used in contrast to the typical energy-method.

e. A more robust and realistic model is used, incorporating both elastic and bulk behavior of the elastomer material (E and K).

All of the above more realistic conditions form the basis of a closed form analytical model to describe the behavior of elastomer seals in the form of contact pressure variation along the seal length, when the seal is subjected to differential fluid pressure. Numerical modeling and simulation of the same problem is also carried out. Conducting experiments on actual swell packers under a variety of oil well conditions is time and cost wise prohibitive, if not practically impossible. Good agreement between the two approaches therefore serves as a good form of validation of the analytical model. A parametric study is then conducted to assess how seal performance is affected by changes in seal design parameters and well conditions (seal geometry, seal compression, differential pressure, and elastomer material) on.

3. Analytical model

When a swell packer comes in contact with a swelling medium (generally water or oil), the elastomer swells primarily in the radial direction. The annular gap is sealed when the elastomer touches the outer casing (or formation). As the rubber swells further, more and more compressive force is generated against the outer casing, causing sealing pressure to increase; **Figure 2**. Normal contact stress is a function of friction at the contact surface. To derive a closed form solution for sealing pressure, a step-wise approach is used. For the initial case of fully bonded contact, the friction coefficient (μ) approaches infinity, resulting in no sliding. In a real well, this can happen when roughness of the rock formation is high enough to prevent sliding. The second scenario is the extreme opposite case of free-sliding contact (zero friction). These two extreme conditions form the basis for the seal pressure model for the more complex but realistic case of frictional contact.

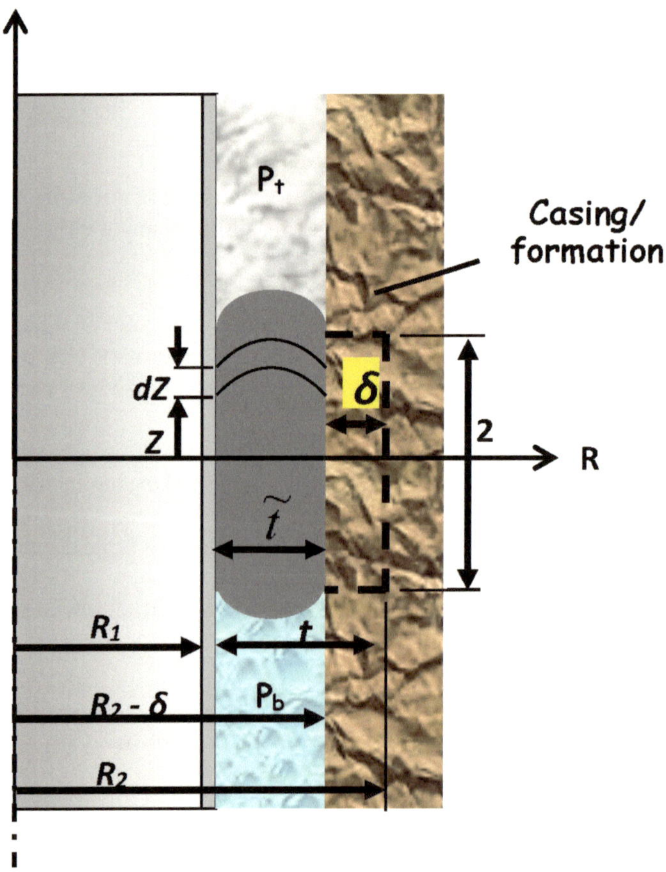

Figure 2.
Geometrical parameters of elastomer seal.

Some of the more fundamental assumptions, closely representing the case of swelling elastomer seals in petroleum drilling operations, are listed below.

a. Elastomer material is homogenous, isotropic, and linear elastic.

b. After deformation, the rubber seal takes on a parabolic shape. In the analytical models developed by Gent and Lindley [16], and Yeoh et al. [17], it is shown that the deformed shape (under compression) does become parabolic.

c. Compression along the circumference of the elastomer seal is uniform. This is necessary to get a valid approximate solution.

d. There is perfect bonding between the steel tubular and the elastomer segment.

e. Friction at the contact surface between the elastomer and the outer casing (or rock formation) follows Coulomb's law.

3.1 Case-1: bonded contact

Forces on an elastomer element, and the change in geometry of this element are shown in **Figure 3**, where h is half seal length, Z is location of differential element along seal axis, dZ is length of differential element, p_t is fluid pressure above the seal (top region), and p_b is fluid pressure below the seal (bottom region). General equation of deformed shape of the seal in the case of the bonded (non-sliding) contact is

$$w(R,Z) = -\frac{4q\,(Z)}{\tilde{t}^2}\left[(R-R_1)^2 - \tilde{t}(R-R_1)\right], \tag{1}$$

where R is seal radius, Z is location of differential element along seal axis, t is seal thickness, R_1 is inner radius of the seal, $\tilde{t}$ is compressed seal thickness, and $q(Z)$ represents the maximum deflection of the compressed seal in the Z-direction. From basic mechanics, the general stress–strain relationships in cylindrical coordinate system are [18].

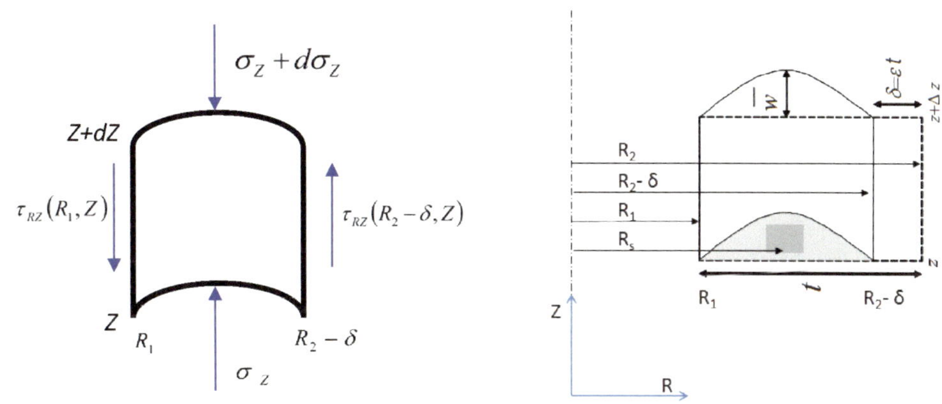

Figure 3.
Forces on an elastomer element (left), and geometry of the volume change element (right).

$$\begin{aligned}
\varepsilon_R &= \frac{\delta}{t} \quad \text{(a)},\\
\varepsilon_Z &= \frac{\partial w}{\partial Z} = -\frac{4}{\tilde{t}^2}\frac{\partial q(Z)}{\partial Z}\left[(R-R_1)^2 - \tilde{t}(R-R_1)\right] \quad \text{(b)},\\
\gamma_{RZ} &= \frac{\partial w}{\partial R} = -\frac{4q(Z)}{\tilde{t}^2}[2(R-R_1)-\tilde{t}] \quad \text{(c)}\\
\sigma_R &= \frac{E}{(1+v)}\varepsilon_R + \frac{3v}{(1+v)}p, \quad \text{(d)}\\
\sigma_Z &= \frac{E}{(1+v)}\varepsilon_Z + \frac{3v}{(1+v)}p, \quad \text{(e)}\\
\tau_{RZ} &= G\gamma_{RZ} = -\frac{4Gq(Z)}{\tilde{t}^2}[2(R-R_1)-\tilde{t}] \quad \text{(f)}
\end{aligned} \tag{2}$$

Here, ε_R, ε_Z, γ_{RZ}, σ_R, σ_Z, and τ_{RZ} are the radial, longitudinal and shear components of strain and stress and respectively, while p is the hydrostatic pressure on the seal ends, and δ is the amount of seal compression. Also, E, K, ν, and G are the elastic modulus, bulk modulus, Poisson's ratio, and shear modulus of the elastomeric material respectively. This set of equations (a to f) will be referred to as Eq. (2). The main objective of the new model is to find the stress in the radial direction (σ_R), which is the sealing pressure created by rubber swelling against outer tubular or formation.

We now apply static force equilibrium ($\sum F_Z = 0$) on the small axi-symmetric elastomer element described in **Figure 3**.

$$\begin{aligned}
&\pi\left[(R_2-\delta)^2 - R_1^2\right]\int_{R_1}^{R_2-\delta}(\sigma_Z + d\sigma_Z - \sigma_Z)dR\\
&+2\pi R_1 dZ\tau_{RZ}(R_1,Z) - 2\pi(R_2-\delta)dZ\tau_{RZ}(R_2-\delta,Z) = 0
\end{aligned} \tag{3}$$

Here, R_2 is the outer radius of the seal after swelling.

Considering seal length to be much larger than its thickness ($H >> t$), a very realistic assumption, and simplifying, we get

$$\frac{dp}{dZ} + \left(\frac{4Eq(Z)}{3v\tilde{t}^3}\right) = 0. \tag{4}$$

Determination of $q(Z)$ is first required to evaluate the pressure $p(Z)$. Using the standard bulk modulus relationship $\Delta V/V_i = -p/K$ as the starting point, we consider the initial volume (V_i) and the change in volume (ΔV) of the differential element, and allow $\Delta Z \to 0$. As we know that $R_s = R_1 + (1/2)\tilde{t}$, we can obtain.

$$\frac{dq\,(Z)}{dZ} = \frac{3R_2\delta}{2R_s\tilde{t}} - \frac{3t(R_2+R_1)}{4R_s\tilde{t}K}p(Z) \tag{5}$$

Eq. (4) now becomes

$$\frac{d^2p}{dZ^2} - \lambda_1{}^2 p(Z) = -\alpha_1, \tag{6}$$

where $\lambda_1{}^2 = \left(\frac{t(R_2+R_1)E}{v\tilde{t}^4 R_s K}\right)$ and $\alpha_1 = \frac{2R_2\delta E}{v\tilde{t}^4 R_s}$.

Solving the above differential equation, we get

$$p(Z) = A\sinh(\lambda_1 Z) + B\cosh(\lambda_1 Z) + \alpha_1/\lambda_1^2. \tag{7}$$

A and B are arbitrary constants determined from boundary conditions. Application of the boundary conditions [$p(Z = h) = p_t$ and $p(Z = -h) = p_b$] and simplification results in the following relations for pressure distribution and normal contact stress in the non-sliding (NS) case:

$$p(Z) = \left(\frac{p_t - p_b}{2\sinh(\lambda_1 h)}\right)\sinh(\lambda_1 Z) + \left(\frac{p_t + p_b - \frac{2\alpha_1}{\lambda_1^2}}{2\cosh(\lambda_1 h)}\right)\cosh(\lambda_1 Z) + \frac{\alpha_1}{\lambda_1^2} \tag{8}$$

$$\sigma_{R\_NS}(Z) = \frac{E}{(1+v)}\frac{\delta}{t} + \frac{3v}{(1+v)} * \left(\left(\frac{p_t - p_b}{2\sinh(\lambda_1 h)}\right)\sinh(\lambda_1 Z) + \left(\frac{p_t + p_b - \frac{2\alpha_1}{\lambda_1^2}}{2\cosh(\lambda_1 h)}\right)\cosh(\lambda_1 Z) + \frac{\alpha_1}{\lambda_1^2}\right). \tag{9}$$

3.2 Case-2: free sliding

A mirror opposite of the first one, the second case represents free sliding or frictionless contact at the elastomer-formation boundary. This translates into zero shear strain (and shear stress) at the contact surface. We again start with force equilibrium, follow a similar series of steps, and use the substitution $R_s = R_1 + (5/8)\tilde{t}$ in Eq. (5) to get.

$$\frac{d^2p}{dZ^2} - \lambda_2^2 p(Z) = -\alpha_2 \tag{10}$$

In the above equation,

$$\lambda_2^2 = \left(\frac{R_1 t(R_2 + R_1)E}{2v(R_2 - \delta + R_1)R_s\tilde{t}^4 K}\right), \text{ and}$$

$$\alpha_2 = \frac{R_1 R_2 \delta E}{v(R_2 - \delta + R_1)R_s\tilde{t}^4}.$$

Further simplification yields the contact pressure and normal stress distribution for the free sliding (S) case:

$$p(Z) = \left(\frac{p_t - p_b}{2\sinh(\lambda_2 h)}\right)\sinh(\lambda_2 Z) + \left(\frac{p_t + p_b - \frac{2\alpha_2}{\lambda_2^2}}{2\cosh(\lambda_2 h)}\right)\cosh(\lambda_2 Z) + \frac{\alpha_2}{\lambda_2^2} \tag{11}$$

$$\sigma_{R\_S}(Z) = \frac{E}{(1+v)}\frac{\delta}{t} + \frac{3v}{(1+v)} * \left(\left(\frac{p_t - p_b}{2\sinh(\lambda_2 h)}\right)\sinh(\lambda_2 Z) + \left(\frac{p_t + p_b - \frac{2\alpha_2}{\lambda_2^2}}{2\cosh(\lambda_2 h)}\right)\cosh(\lambda_2 Z) + \frac{\alpha_2}{\lambda_2^2} \right). \tag{12}$$

3.3 Case-3: frictional contact

The last and final case is the more realistic one of frictional contact at the elastomer-formation boundary; somewhere in between the two extreme cases of bonded contact and free-sliding. Frictional contact is defined at a cylindrical surface rather than a flat surface. If the rubber is vulcanized to the inner casing (fully bonded on one side), and contact on the other side (formation or outer casing) is frictional in nature, there will be a non-symmetrical bulging in the middle. Its magnitude is dictated by friction at outer contact surface; as friction coefficient (μ) increases, the amount of sliding decreases. To derive the normal contact stress distribution, we consider a weighted sum of the two extreme cases described above.

$$\sigma_{R\_F}(Z) = (1 - \mu)[\sigma_{R\_S}(Z) + (\mu)[\sigma_{R\_NS}(Z)] \tag{13}$$

Expanded form of this equation is shown below.

$$\sigma_{R\_F}(Z) = \left(\begin{array}{l} \left[\left(\frac{p_t - p_b}{2\sinh(\lambda_2 h)}\right)\sinh(\lambda_2 Z) + \left(\frac{p_t + p_b - \frac{2\alpha_2}{\lambda_2^2}}{2\cosh(\lambda_2 h)}\right)\cosh(\lambda_2 Z) + \frac{\alpha_2}{\lambda_2^2} \right] \\ -\frac{3v\mu}{(1+v)} \left[\begin{array}{l} \left(\left(\frac{p_t - p_b}{2\sinh(\lambda_2 h)}\right)\sinh(\lambda_2 Z) + \left(\frac{p_t + p_b - \frac{2\alpha_2}{\lambda_2^2}}{2\cosh(\lambda_2 h)}\right)\cosh(\lambda_2 Z) + \frac{\alpha_2}{\lambda_2^2} \right) \\ -\left(\left(\frac{p_t - p_b}{2\sinh(\lambda_1 h)}\right)\sinh(\lambda_1 Z) + \left(\frac{p_t + p_b - \frac{2\alpha_1}{\lambda_1^2}}{2\cosh(\lambda_1 h)}\right)\cosh(\lambda_1 Z) + \frac{\alpha_1}{\lambda_1^2} \right) \end{array} \right] \end{array} \right). \tag{14}$$

Figure 4 shows the variation of contact pressure along the seal length for the three cases, described by this last Eq. (14). Making intuitive sense, pressure curve for this frictional contact lies between the previous two extreme cases of full bonding and free sliding. Also, higher friction generates higher sealing pressure. As can be seen in **Figure 4**, location of maximum pressure is at the seal center. This value can be determined by putting $Z = 0$ in Eq. (14), yielding the following equation.

$$\sigma_{R\_F}(Z=0) = \left(\left[\left(\frac{p_t + p_b - \frac{2\alpha_2}{\lambda_2^2}}{2\cosh(\lambda_2 h)}\right) + \frac{\alpha_2}{\lambda_2^2} \right] - \frac{3v\mu}{(1+v)} \left[\begin{array}{l} \left(\left(\frac{p_t + p_b - \frac{2\alpha_2}{\lambda_2^2}}{2\cosh(\lambda_2 h)}\right) + \frac{\alpha_2}{\lambda_2^2} \right) \\ -\left(\left(\frac{p_t + p_b - \frac{2\alpha_1}{\lambda_1^2}}{2\cosh(\lambda_1 h)}\right) + \frac{\alpha_1}{\lambda_1^2} \right) \end{array} \right] \right) \tag{15}$$

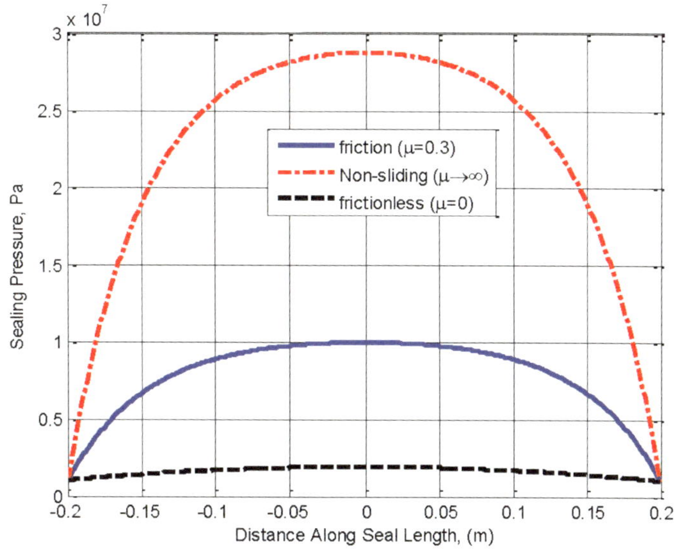

Figure 4.
Normal contact stress distribution along the elastomer seal for the three cases (sliding, no-sliding, frictional); $E = 0.20$ *MPa*, $p_t = 1$ *MPa*, $p_b = 1$ *MPa*, $t = 20$ *mm*, $H = 400$ *mm*, $R_1 = 100.33$ *mm*, $\delta = 1$ *mm.*

4. Numerical model

Numerical modeling and simulation of swelling-elastomer seal has been done here using the commercial finite element analysis (FEA) package ABAQUS. An axisymmetric model was deemed sufficient as both loads and geometry are

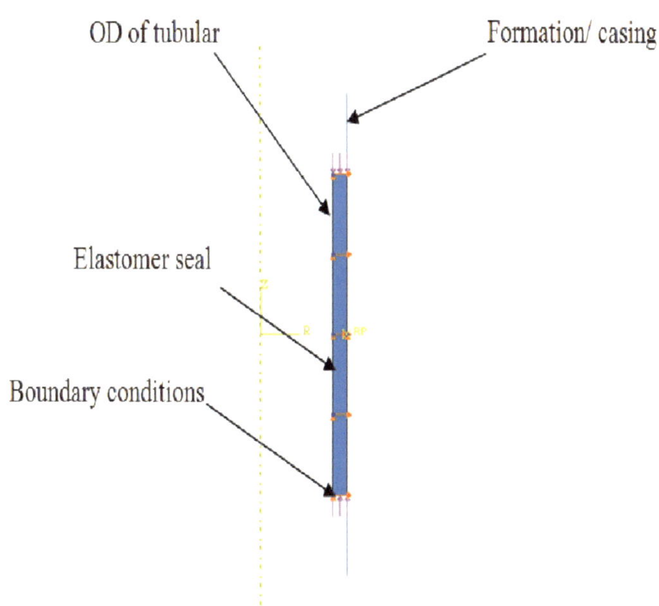

Figure 5.
Finite element model showing the elastomer seal and boundary conditions.

symmetric. The rubber element is treated as a deformable material, and in comparison the outer casing or formation is treated as a rigid body. Using an actual regional oilfield as a base, numerical model was developed for a seal length (H) of 400 mm; a steel tubular having an outer radius of (R_1) of 100.33 mm, same as the seal inner radius; and an elastomer seal having a thickness (t) of 20 mm. By using a Poisson's ratio (v) value of 0.4999, the seal material was rendered almost incompressible for bulk deformation. The neo-Hookean hyperelastic material model was used to represent the elastomer material. Actual data sets from compression experiments conducted on laboratory size elastomer samples at various swelling intervals were used to extract the model coefficients. **Figure 5** is a depiction of the FE model with axisymmetric boundary conditions, while **Figure 6** shows the simulation results for variation of sealing pressure along the length of the seal.

Analytical and FEA results for seal pressure variation over the seal length are compared against each other in **Figure 7** for two values of μ. Simulated and theoretical results were plotted for many other conditions also, but are not shown here due to space limitation. All the comparative graphs exhibit a close agreement

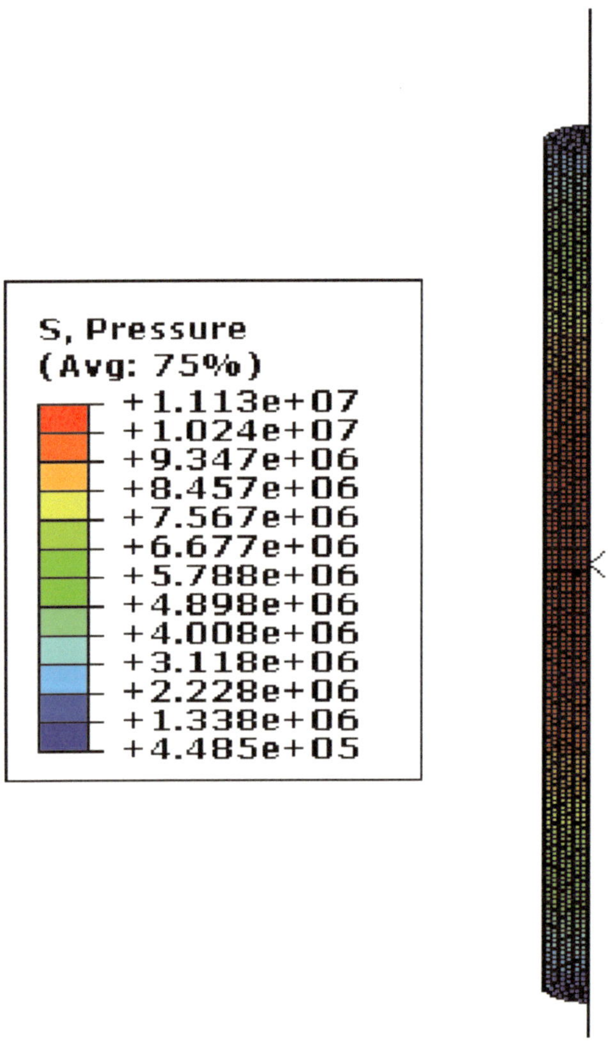

Figure 6.
Finite element simulation of variation of sealing pressure along the elastomer seal.

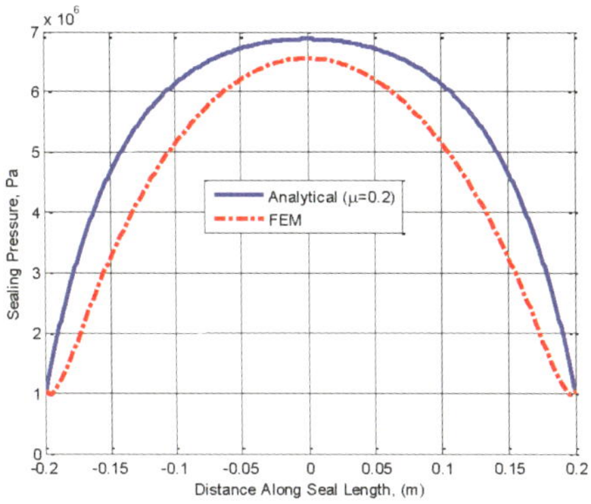

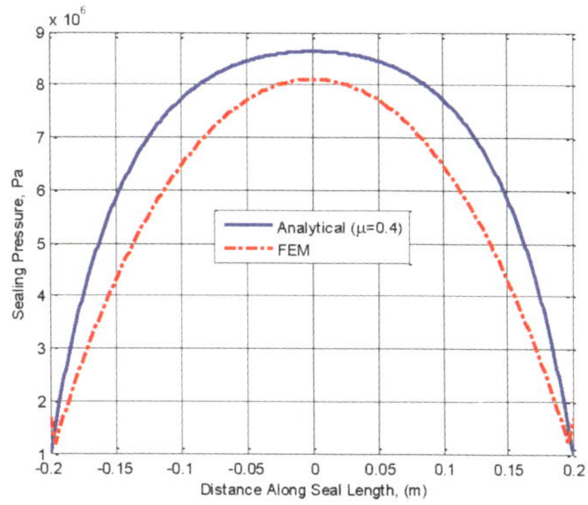

Figure 7.
Comparison of analytical and numerical results for sealing pressure distribution for $\mu = 0.2$ *(left) and* $\mu = 0.4$ *(right);* $E = 0.20$ *MPa,* $p_t = 1$ *MPa,* $p_b = 1$ *MPa,* $t_i = 20$ *mm,* $H = 400$ *mm,* $R_1 = 100.33$ *mm,* $\delta = 1$ *mm.*

between FE and analytical results. However, it is noteworthy that the analytical model gives a slightly higher sealing pressure than the FE model in all cases. A probable explanation is that the numerical model also includes material nonlinearity (neo-Hookean hyperelastic material model), while the analytical model considers the rubber to be a linear elastic material.

5. Results and discussion

The last Eq. (15) describes the contact pressure generated by the elastomer seal as a function of seal compression, seal length, seal thickness, and elastomer material properties. Given below is a parametric study of how the maximum sealing pressure behaves as different seal parameters are varied.

Maximum contact pressure (p_{max}) is plotted as a function of seal length (H) for different values of seal compression (δ) in **Figure 8**. Sealing pressure distribution along the seal is not constant but varies nonlinearly depending on seal parameters and loading conditions, with maximum sealing pressure occurring at the center of the seal length. In all cases, there is an initial sharp increase in sealing pressure for increasing seal length, which gradually becomes almost constant once the seal length crosses a value of 40 cm. This observation has a major practical and economic bearing for field engineers. ***A seal length of more than 40 cm is impractical; it will not generate higher seal pressure. Limiting length of individual rubber elements to 40 cm can result in major cost and weight savings.*** To be practically more effective, several 40-cm seal elements can be mounted in series for more effective zonal isolation.

It can be observed from **Figure 8** that larger amount of seal compression produces higher pressure curves. Again, this is as expected; more compression generates higher sealing pressure. Elastomer can be compressed by a larger amount by using a material that swells more under the given conditions, or by using swell packers together with solid expandable tubular (SET) technology. In an SET application, a petroleum tubular (with elastomer seal mounted on it) is expanded downhole by forcing a conical mandrel through it, thus blocking the perforations and shutting off an unwanted zone.

Variation of maximum seal pressure as a function of seal length can be seen in **Figure 9** for different values of seal thickness. The apparently counter-intuitive observation, that sealing pressure decreases with increasing seal thickness, needs a little explanation. As explained earlier in Chapter-7, this happens because seal pressurization is modeled here (numerically) in the form of a fixed amount of displacement, while seal thickness is changed. At the beginning, the elastomer seal is just touching the casing wall. As the displacement (δ) is now applied, the elastomer undergoes compression. If the seal is thin, the compression ratio (δ divided by seal thickness t) is much larger, thus the behavior seen in **Figure 9**. It is therefore better to use the term "compression ratio" rather than "seal compression." Sealing pressure increases (almost proportionately) as compression ratio is increased, as expected; **Figure 10**.

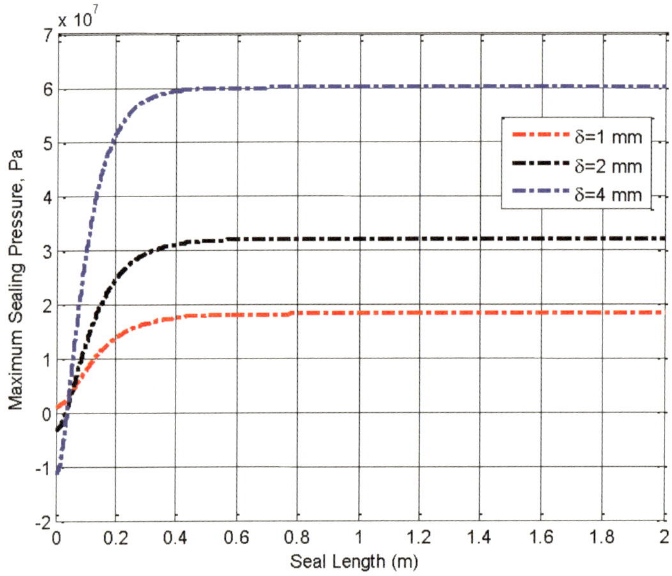

Figure 8.
Maximum seal pressure as a function of seal length for different values of seal compression; E = *0.33 MPa,* p_t = *1 MPa,* p_b = *0 MPa,* t = *20 mm.*

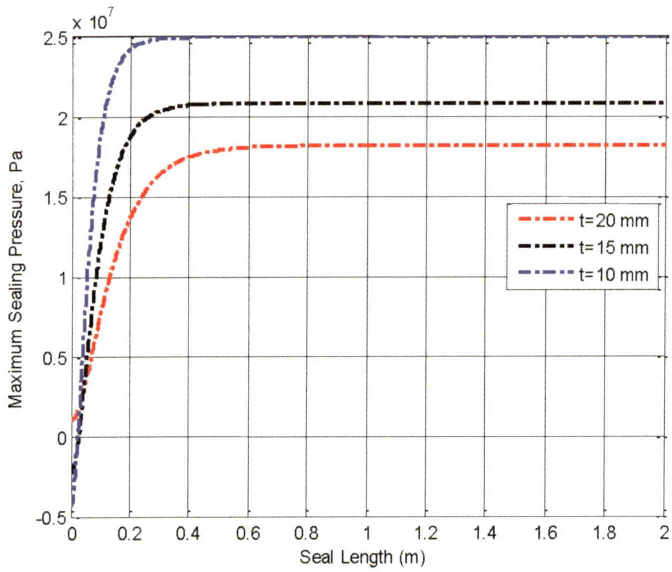

Figure 9.
Maximum seal pressure as a function of seal length for different values of seal thickness; E = *0.33 MPa,* p_t = *1 MPa,* p_b = *0 MPa,* H = *400 mm,* R_1 = *100.33 mm,* μ = *0.2.*

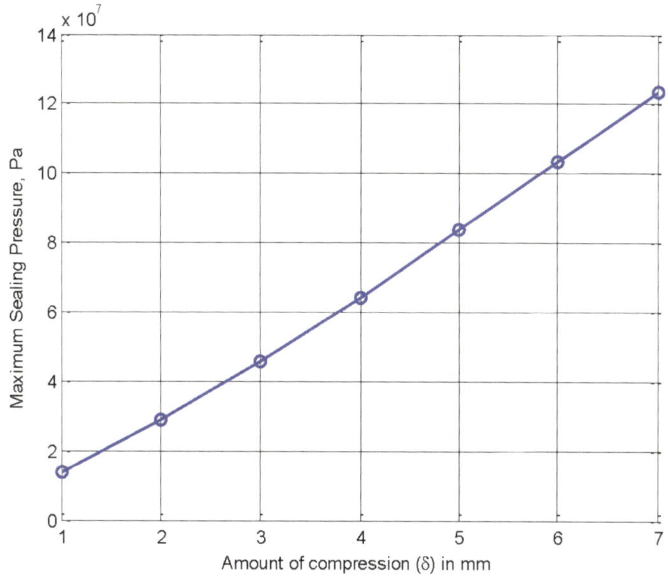

Figure 10.
Maximum sealing pressure is directly proportional to seal compression for constant seal thickness; $E = 0.33\ MPa$, $p_t = 1\ MPa$, $p_b = 0\ MPa$, $t = 20\ mm$, $H = 400\ mm$, $R_1 = 100.33\ mm$, *and* $\mu = 0.2$.

6. Conclusions

An analytical (mathematical) model has been developed to quantify the pressure distribution along an elastomer seal as a function of seal geometry (thickness, length, etc.), material properties of rubber and steel pipe (E, G, K, and v), friction at the contact surface, and differential pressure across the seal in an oil well. This approach of using actual swelling elastomer seals and packers as a base makes this model significantly different from previous studies. On the inner side, the elastomer is bonded (vulcanized) to a petroleum tubular, and on the outer side it has a frictional contact with an outer casing or rock formation. Instead of considering only the elastic modulus of the material, the model uses both E and K values of the elastomer material. Rather than assuming the elastomer to be incompressible, the value of Poisson's ratio is taken to be different from 0.5. As opposed to the conventional energy-method, a volume-change approach is adopted.

The problem is also modeled and simulated numerically. Results from numerical simulations and analytical model are quite close to each other, confirming that the mathematical model gives good prediction of sealing behavior of the elastomer. Variation of sealing pressure along the seal is nonlinearly in nature, and depends on seal parameters and well conditions. It reaches a maximum value at the center of the seal. Very longer seals are impractical; after a seal length of 40 cm, sealing pressure does not increase much. Larger compression ratio generates higher sealing pressure. More seal compression can be achieved by using an elastomer that swells more, or by using swelling elastomers together with SET expansion. If the applied seal compression is kept constant, thinner seal yields higher compression ratio; this generates higher sealing pressure.

This analytical model is a major improvement on previous models, saves a lot of computational time in comparison with numerical models, and gives reasonably accurate prediction of elastomer seal performance under various actual oil and gas field conditions. The obtained results are of major interest for academics and researchers, petroleum engineers, and swell packer designers and manufacturers.

Author details

Sayyad Zahid Qamar[1]*, Maaz Akhtar[2] and Tasneem Pervez[1]

1 Mechanical and Industrial Engineering Department, Sultan Qaboos University, Muscat, Oman

2 Mechanical Engineering Department, N.E.D. University of Engineering and Technology, Karachi, Pakistan

*Address all correspondence to: sayyad@squ.edu.om

References

[1] Bellarby J (2009), Well Completion Design, Elsevier

[2] Kennedy G, Lawless A, Shaikh K, Alabi T (2005) "The Use of Swell Packers as a Replacement and Alternative to Cementing," Society of Petroleum Engineers, SPE Paper # 95713, *SPE Annual Technical Conference and Exhibition*, 9-12 October 2005, Dallas, Texas

[3] Daniel L, Hertz J. "Chapter-1: Introduction" in AN Gent (ed) Engineering With Rubber: How to Design Rubber Components. 2nd ed. Hanser-Gardner: Publishers; 2000

[4] Marckmann G, Verron E. Comparison of Hyperelastic Models for Rubber-Like Materials. Rubber Chemistry and Technology. 2006;**79**(5): 835-858

[5] Al-Yami AS, Al-Arfaj MK, Nasr-el-Din HA, Al-Saleh SH, Al-Humaidi AS, Al-Moajil AM (2008b) "Lab Investigation of Oil Swelling Elastomers for Smart Well Completions," SPE Paper #113135, *Europec/EAGE Conference and Exhibition*, 9-12 June 2008, Rome, Italy

[6] Qamar SZ, Pervez T, Akhtar M (2016) "Performance Evaluation of Water-Swelling and Oil-Swelling Elastomers," Journal of Elastomers and Plastics, 48 (6) 2016, p 535-545

[7] Qamar SZ, Pervez T, Akhtar M, Al-Kharusi MSM. Design and Manufacture of Swell Packers: Influence of Material Behavior. Materials and Manufacturing Processes. 2012;**27**(7):727-732

[8] Qamar SZ, Hiddabi SA, Pervez T, Marketz F. Mechanical Testing and Characterization of a Swelling Elastomer. Journal of Elastomers and Plastics. 2009;**41**(5):415-431

[9] Al-Yami AS, Nasr-el-Din HA, Al-Humaidi AS (2008a) "Investigation of Water Swelling Elastomers: Advantages, Limitations, and Recommendations," SPE Paper # 114810, *SPE Annual Technical Conference and Exhibition*, 21-24 September 2008, Denver, Colorado, USA

[10] Akhtar M, Qamar SZ, Pervez T, Al-Jahwari FK (2018) "Performance Analysis of Swelling Elastomer Seals," Petroleum Science and Engineering, 165 (2018), p 127-135

[11] Qamar SZ, Akhtar M, Al-Kharusi MSM (2013) "Effect of Swelling Behavior on Elastomeric Materials: Experimental and Numerical Investigation," ASME-2013 International Mechanical Engineering Congress and Exhibition (IMECE-2013), 15-21 November 2013, San Diego, California

[12] Al-Kharusi MSM, Qamar SZ, Pervez T, Akhtar M (2011) "Non-Linear Model for Evaluation of Elastomer Seals Subjected to Differential Pressure," Paper # SPE-149032, SPE/DGS Saudi Arabia Section Technical Symposium and Exhibition, 15-18 May 2011, Al-Khobar, Saudi Arabia

[13] Al-Kharusi MSM, Qamar SZ, Pervez T, Akhtar M (2013) "Elastomer Seal with Frictional Contact — Analytical Solution," ASME-2013 International Mechanical Engineering Congress and Exhibition (IMECE-2013), 15-21 November 2013, San Diego, California

[14] Gent AN, Meinecke EA. Compression, Bending, and Shear of Bonded Rubber Blocks. Polymer Engineering and Science. 1970;**10**(1): 48-53

[15] Chalhoub MS, Kelly JM. Analysis of Infinite-Strip-Shaped Base Isolator with

Elastomer Bulk Compression. Journal of Engineering Mechanics. 1991;**117**(8): 1791-1805

[16] Gent AN, Lindley PB. The Compression of Bonded Rubber Blocks. Proceedings of the Institution of Mechanical Engineers. 1959;**173**(p): 111-122

[17] Yeoh OH, Pinter GA, Banks HT. Compression of Bonded Rubber Blocks. *Rubber Chemistry and Technology*, Vol. 2002;**75**:549-561

[18] Beer FP, Johnston ER, Dewolf JT, Mazurek D. Mechanics of Materials. 7th ed. McGraw-Hill: Publications; 2014

Chapter 11

New Analytical Model for Swellable Materials

Sayyad Zahid Qamar, Maaz Akhtar and Tasneem Pervez

The history of science shows that theories are perishable. With every new truth that is revealed we get a better understanding of Nature, and our conceptions and views are modified.

Nikola Tesla

Abstract

As discussed in Chapter 6, numerical prediction of swelling can be attempted using existing hyperelastic material models available in commercial finite element (FE) packages. However, none of these models can accurately represent the behavior of swelling elastomers. The major shortcoming of currently available swelling models is that they consider Gaussian statistics for mechanical contribution of configuration entropy, which is based on chains having limited extensibility. Some later models (not yet incorporated into commercial FE packages) can give a reasonable account of certain behavior patterns in swelling elastomers, but do not explain other aspects well. One of the new approaches is to treat swelling elastomers as gels. As described earlier, gels are mostly liquid, yet they behave like solids due to a three-dimensional cross-linked network within the liquid. Many authors consider gel as poro-elastic or porous and use Darcy's law to model the amount of fluid influx. However, a swollen elastomer mostly consists of the solvent. When an external load is applied, maximum resistance comes from the solvent molecules as in diffusion. Also, most of the new models are quite complex in concept and formulation, and there is a serious need for a scientifically simpler model.

Keywords: swelling elastomer, new material model, continuum mechanics, non-Gaussian statistics

1. Introduction

As discussed in Chapter 6, numerical prediction of swelling can be attempted using existing hyperelastic material models available in commercial finite element (FE) packages. However, none of these models can accurately represent the behavior of swelling elastomers [1]. The major shortcoming of currently available swelling models is that they consider Gaussian statistics for mechanical contribution of configuration entropy, which is based on chains having limited extensibility [2]. These models assume small stretch of elastomer chains, while swelling elastomers experience much larger stretches. That is why they show only modest agreement with experimental data.

Some later models (not yet incorporated into commercial FE packages) can give a reasonable account of certain behavior patterns in swelling elastomers, but do not

IntechOpen

explain other aspects well [3]. One of the new approaches is to treat swelling elastomers as gels. As described earlier, gels are mostly liquid, yet they behave like solids due to a three-dimensional cross-linked network within the liquid. Many authors consider gel as poro-elastic or porous and use Darcy's law to model the amount of fluid influx. However, a swollen elastomer mostly consists of the solvent. When an external load is applied, maximum resistance comes from the solvent molecules as in diffusion. Also, most of the new models are quite complex in concept and formulation, and there is a serious need for a scientifically simpler model.

2. Proposed model: salient features

Presented below is the development of a new material model for the prediction of swelling in elastomeric materials, using a continuum mechanics approach. To account for the changes in configuration entropy of the elastomer chains due to swelling, almost all available models [4–7] use classical Gaussian-statistics in which chains are considered to have limited extensibility [8]. Swelling elastomers undergo large deformations as the chain network stretches more. More realistic non-Gaussian statistics is therefore used for model development here, to account for large mechanical stretches, in terms of mechanical contribution of configuration entropy.

It was concluded in Chapter 6 that Ogden model gives the closest predictions for swelling elastomers [1]. The hyperelastic portion of the new model is therefore based on the phenomenological stretch-based Ogden model. This second-order non-Gaussian strain energy function is used to define changes due to configuration entropy [9]. Rather than treating swelling as an osmosis problem, diffusion is considered to be the mechanism responsible for fluid influx. Flory-Huggins theory is used for incorporating the thermodynamics of mixing of polymer and solvent (absorption of fluid into the elastomer). Unlike other models, which consider only some of the pertinent parameters, proposed model includes most of the relevant material and structural properties of the elastomer, and environmental conditions (temperature, water salinity, coefficient of diffusion, polymer-solvent interaction parameter, etc).

The formulation is based on thermodynamically consistent diffusion-deformation theory for elastomer gels considering the interaction and mixing of polymer and solvent. Solid and liquid like chemical species are considered as a single homogenized continuum [7, 10]. It is assumed that Helmholtz free energy can be divided into network stretching and Fluid-polymer mixing [11]. Flory-Huggins theory is used to describe the changes in entropy due to solid–fluid mixing.

Fluid imbibition in swelling elastomers follows the mechanism of diffusion [4, 12]. Diffusion equations are formulated through suitable balance laws for fluid content. Coefficient of diffusion (D) for fluid molecules is assumed to be isotropic and independent of deformation gradient. As fluid molecules form a majority portion of the swollen elastomer, this simplification seems quite realistic. Deformation due to swelling may occur in two ways, short range or long range, as shown in **Figure 1**. Initially, the solvent molecules diffuse and re-arrange so that shape changes but volume remains constant. In long-range motion, gel changes shape as well as volume. It is assumed in developing the model that amount of solvent only transports when change of volume occurs, and it remains constant when elastomer undergoes only change of shape.

Swelling elastomers are assumed to be incompressible because they possesses high bulk modulus and low shear modulus [13]. Change in volume is negligible as

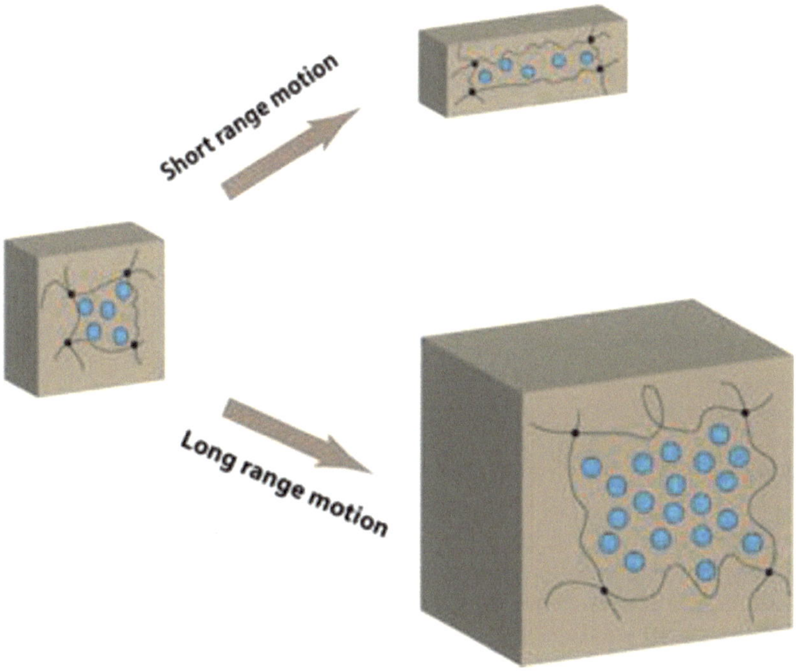

Figure 1.
Deformation under long and short range motion of gel.

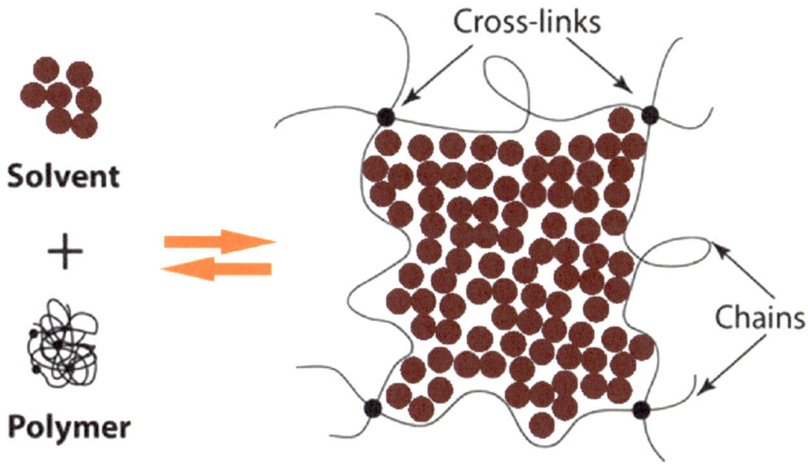

Figure 2.
Imbibition of solvent and stretching of chain.

compared to shape change. Gels are soft materials that can deform easily and can undergo volume changes equal to several times its initial volume. This change in volume occurs only as a result of imbibition of solvent molecules. It should be noted that volume of swollen elastomer is the state bounded by material points that deform with the elastomer chains. Swelling of gel actually means deformation of the elastomer network. When a swellable elastomer is placed in a solvent, it absorbs it, resulting in stretching of chains with increase in volume; **Figure 2**. This unique feature necessitates the consideration of large deformation and helps in developing the mathematical basis for mechanics of swelling. Mathematical structure used to derive deformation of elastomer network due to swelling is similar to the approach used in rubber elasticity. Rubber can be treated as a special case of swelling elastomer with no fluid. Succeeding sections describe the continuum mechanics theory for swelling elastomers.

3. Proposed model: mathematical formulation

The new model for swelling in elastomers is developed in this section, describing the relevant contributions of kinematics, force equilibrium, solvent equilibrium, system free energy, kinematic constraint for network incompressibility, kinetics, and thermodynamics of mixing.

3.1 Kinematics

Polymer-fluid mixture is considered to be a single homogenized continuum body permitting diffusion of solvent. Reference configuration is a three dimensional dry state of elastomer represented by Ω_0 contained within a surface denoted by S_0. Current configuration represents the deformed swollen state of elastomer network. Intermediate or auxiliary state is the local distortion of elastomer because of volume increase only. Later rotation and stretching of swollen elastomer gives the mechanical elastic deformation in final configuration.

Consider a particle A labeled by its position $\mathbf{X}$ in the reference state. Deformation of entire material is described by motion and solvent concentration with smooth mapping of all material particles to the final state. After deformation, particle A attains a new position $\mathbf{x}$ in current configuration as shown in **Figure 3**. Similarly, each particle is deformed and attains a new position in the final sate.

Analytical description of the deformation of a continuum can follow either *Eulerian* (spatial) or *Lagrangian* (material) descriptions. In *Eulerian* approach, deformation is referred to the current state, while in *Lagrangian* description deformation is referred to reference configuration [14]. In continuum mechanics of solid bodies, spatial description is less beneficial since current configuration is not known. Consider particles A and B in Ref. configuration represented by positions $\mathbf{X}_A$ and $\mathbf{X}_B$, respectively. Elemental vector is given by

$$d\mathbf{X} = \mathbf{X}_B - \mathbf{X}_A. \qquad (1)$$

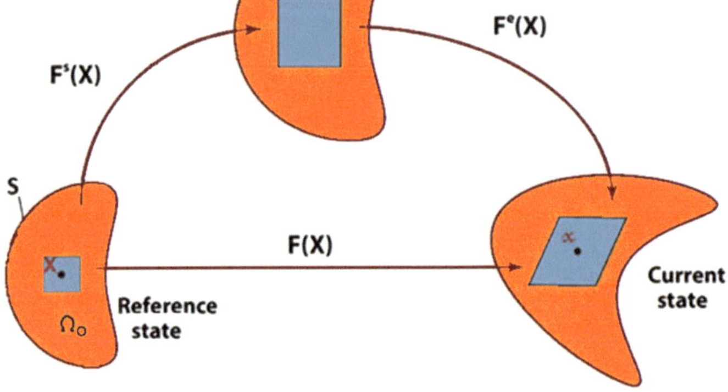

Figure 3.
Schematic of the reference (dry), intermediate (swollen), and current (deformed) configurations.

After deformation at time $(\mathbf{t})$, particles occupy spatial positions $\mathbf{x}_A$ and $\mathbf{x}_B$ in current configuration, having elemental vector given by

$$\mathrm{dx} = \mathbf{x}_\mathrm{B} - \mathbf{x}_\mathrm{A}. \tag{2}$$

Displacement of the particles A and B is given by

$$\mathbf{u}_\mathrm{A} = \mathbf{x}_\mathrm{A} - \mathbf{X}_\mathrm{A}, \qquad \mathbf{u}_\mathrm{B} = \mathbf{x}_\mathrm{B} - \mathbf{X}_\mathrm{B}. \tag{3}$$

Displacement attained by material in the current configuration can be written as

$$\mathbf{u}(\mathbf{X}, \mathrm{t}) = \mathbf{x} - \mathbf{X}. \tag{4}$$

Relationship between current and reference state (before and after deformation) is expressed by the *deformation gradient tensor* $\boldsymbol{F}(\boldsymbol{X})$, expressed as [14].

$$\begin{aligned} d\mathbf{x} &= \mathbf{F}\cdot d\mathbf{X} = d\mathbf{X}\cdot \mathbf{F}^T \\ \mathbf{F} &= \left(\frac{\partial \mathbf{x}}{\partial \mathbf{X}}\right) \text{ or } \mathbf{F}_{ij} = \frac{\partial \mathbf{x}_i(\mathbf{X}, t)}{\partial \mathbf{X}_j}. \end{aligned} \tag{5}$$

A crucial kinematic constituent of the current model is the multiplicative decomposition of the deformation gradient into swelling and mechanical elastic parts as suggested by Flory [11]. Local distortion of the material at X due to swelling is given by $\boldsymbol{F}^s(\boldsymbol{X})$, while rotation and stretching of swollen network is specified by $\boldsymbol{F}^e(\boldsymbol{X})$. Hence, deformation gradient can be written as

$$\mathbf{F} = \mathbf{F}^e\mathbf{F}^s. \tag{6}$$

Jacobian $\boldsymbol{J}$ is the determinant of deformation gradient such that

$$\begin{aligned} &\mathrm{J} = \det\mathbf{F} > 0. \\ &\mathrm{J}^\mathrm{e} = \det\mathbf{F}^\mathrm{e} \quad \text{and} \quad \mathrm{J}^\mathrm{s} = \det\mathbf{F}^\mathrm{s} \end{aligned} \tag{7}$$

Using Eq. (6) and definition of Jacobin from Eq. (7), we can write

$$\mathrm{J} = \mathrm{J}^\mathrm{e}\mathrm{J}^\mathrm{s}. \tag{8}$$

3.2 Force equilibrium

The principle of virtual work is considered to be a fundamental law in continuum mechanics [15]. It states that the magnitude of virtual work for the forces acting on a particle in equilibrium is zero for any arbitrary virtual motion. The balance of body forces $(\mathbf{b})$ per unit volume at equilibrium in reference configuration is given by

$$\mathrm{div}\mathbf{s} + \mathbf{b} = 0, \tag{9}$$

where $\mathbf{s}$ is a statically admissible stress field that holds true for any test function ξ. Multiplying Eq. (9) by this test function and integrating with respect to volume gives

$$\begin{aligned} &\int_\mathrm{V} (\mathrm{div}\mathbf{s} + \mathbf{b})\xi \mathrm{dV} = 0 \\ &\Rightarrow \int_V \xi \, \mathrm{div}\mathbf{s} dV + \int_V \mathbf{b}\, \xi dV = 0. \end{aligned} \tag{10}$$

We know that

$$\mathrm{div}(A\varepsilon) = \varepsilon\,\mathrm{div}A^{T} + \mathrm{tr}(A\,\mathrm{grad}\varepsilon), \text{and } tr\left(\mathbf{A}^{T}\mathbf{B}\right) = \mathbf{A}:\mathbf{B}.$$

Eq. (10) thus becomes

$$\int_{V}\{\mathrm{div}(\mathbf{s}\xi) - \mathbf{s} : \mathrm{grad}(\xi) + \mathbf{b}\cdot\xi\}dV = 0. \tag{11}$$

Applying Gauss divergence theorem to first term of Eq. (11) transforms the behavior from inside the body to the vector field ($\mathbf{n}$) through the surface,

$$\int_{V}\mathrm{div}(\mathbf{s}\xi)dV = \int_{S}\mathbf{sn}\xi\,dA.. \tag{12}$$

Cauchy's law states that a stress tensor exists which maps the normal to a surface into traction vector acting on the surface. When applied on Eq. (12) traction boundary loads ($\mathbf{t}$) per unit area are added to the equation. Rearranging the terms gives the statement for principle of virtual work:

$$\int_{V}\mathrm{s}_{ij}\frac{\partial\xi_{i}}{\partial\mathbf{X}_{j}}\mathrm{dV} = \int_{V}\mathbf{b}_{i}\xi_{i}\,\mathrm{dV} + \int_{S}\mathbf{t}_{i}\xi_{i}\,\mathrm{dA} = 0. \tag{13}$$

3.3 Solvent equilibrium

Changes in solvent concentration for any material are typically explained through diffusion across the boundary. Let $C(\mathbf{X},t)$ denote the concentration of solvent molecules absorbed by the polymer per unit volume of reference configuration, with the assumption that no chemical reactions occur. Solvent flux ($\boldsymbol{\phi}$) per unit area represents the amount of solvent particles entering through the boundary per unit time. For all the particles, conservation of solvent molecules takes the following form [10].

$$\int_{V}\dot{\mathrm{C}}\mathrm{dV} = -\int_{S}\boldsymbol{\phi}\cdot\mathbf{m}\,\mathrm{dA}. \tag{14}$$

Applying the divergence theorem to Eq. (14), and rearranging the terms, yields the following balance law for solvent content

$$\int_{V}(\dot{\mathrm{C}} + \mathrm{div}\boldsymbol{\phi})dV = 0. \tag{15}$$

For this equation to hold, we must have

$$\begin{aligned}&\dot{\mathrm{C}} = -\mathrm{div}\boldsymbol{\phi}\\ \text{or}\quad &\frac{\partial\mathrm{C}(\mathbf{X},\mathrm{t})}{\partial\mathrm{t}} + \frac{\partial\boldsymbol{\phi}_{\mathrm{j}}(\mathbf{X},\mathrm{t})}{\partial\mathbf{X}_{\mathrm{j}}} = 0.\end{aligned} \tag{16}$$

The equation for solvent diffusion at the rate (ψ) across the boundary at flux ($\boldsymbol{\phi}$) is formulated through the law of balance law [5] given by

$$\begin{aligned}&\psi = -\boldsymbol{\phi}\cdot\boldsymbol{m}\\ &\Rightarrow\ -\boldsymbol{\phi}_{j}(\boldsymbol{X},t)\boldsymbol{m}_{j}(\boldsymbol{X},t) = \psi(\boldsymbol{X},t).\end{aligned} \tag{17}$$

Multiplying Eq. (16) with an arbitrary test function ζ and integraing with respect to volume gives

$$\int_V \frac{\partial C}{\partial t}\zeta dV + \int_V \frac{\partial \phi_j}{\partial \mathbf{X}_j}\zeta dV = 0. \tag{18}$$

Similarly, multiplying Eq. (17) with the same arbitrary test function (ζ), and integrating with respect to area, we get

$$-\int_S \boldsymbol{\phi}_j \boldsymbol{m}_j \zeta dA = \int_S \psi \zeta dA. \tag{19}$$

Applying divergence theorem on Eq. (18), putting Eq. (19) in the resultant, and on simplification we get

$$\int_V \frac{\partial C}{\partial t}\zeta dV = \int_V \boldsymbol{\phi}_j \frac{\partial \zeta}{\partial \mathbf{X}_j} dV + \int_S \psi \zeta dA. \tag{20}$$

Chemical potential gradient is required to induce swelling in the polymer. Solvent molecules in dilute phase have higher chemical potential than those in concentrated phase. This difference in potential causes solvent flow from solution to polymer. Chemical potential ($\boldsymbol{\mu}$) is characterized by the energy flow(Θ) due to solvent transport across the boundary as discussed by Gurtin et al. [16], and is given by

$$\begin{aligned} \Theta &= -\int_S \boldsymbol{\mu\phi} \cdot \boldsymbol{m} dA \\ \Rightarrow \Theta &= -\int_S \boldsymbol{\mu} \psi dA. \end{aligned} \tag{21}$$

3.4 Free energy of the system

Free energy density is a function of concentration and deformation gradient. For an element of volume dV, Helmholtz free energy of the swelling elastomer in the current configuration is denoted by WdV. When equilibrium is attained between the fluid and the swollen elastomer, chemical potential becomes homogenous both inside the elastomer and in the surrounding solvent. Small changes δC in the fluid concentration and $\delta \mathbf{F}$ in the deformation gradient of the elastomer causes changes in the free-energy density δW.

$$\delta W = \frac{\partial W(\mathbf{F}, C)}{\partial \mathbf{F}} \delta \mathbf{F} + \frac{\partial W(\mathbf{F}, C)}{\partial C} \delta C. \tag{22}$$

At equilibrium, free energy of the system is the amount of work done by all the loads and due to solvent transport. For small changes, rate of change of free energy of the system (Λ) is given by

$$\frac{\delta \Lambda}{\delta t} = \int_V \frac{\delta W}{\delta t} dV - \int_V \boldsymbol{b} \frac{\delta x}{\delta t} dV - \int_S \boldsymbol{t} \frac{\delta x}{\delta t} dA - \int_S \mu \psi dA. \tag{23}$$

Replacing test function with $(\delta x / \delta t)$ in Eq. (13), we get

$$\int_V \boldsymbol{b} \frac{\delta x}{\delta t} dV + \int_S \boldsymbol{t} \frac{\delta x}{\delta t} dA = \int_V \mathbf{s} \frac{\partial}{\partial \mathbf{X}} \left(\frac{\delta x}{\delta t} \right) dV = \int_V \mathbf{s} \left(\frac{\delta \mathbf{F}}{\delta t} \right) dV. \tag{24}$$

Dividing Eq. (22) by δt and placing in Eq. (20), we get the following equation:

$$\frac{\delta W}{\delta t} = \frac{\partial W}{\partial \mathbf{F}}\left(\frac{\delta \mathbf{F}}{\delta t}\right) + \frac{\partial W}{\partial C}\left(\int_V \boldsymbol{\phi}\frac{\partial}{\partial \mathbf{X}} dV + \int_S \psi dA\right). \tag{25}$$

Substituting Eqs. (24) and (25) in Eq. (23) and rearranging the terms, we get

$$\frac{\delta \Lambda}{\delta t} = \int_V \left(\frac{\partial W}{\partial \mathbf{F}} - \mathbf{s}\right)\frac{\delta \mathbf{F}}{\delta t} dV + \int_S \left(\frac{\partial W}{\partial C} - \mu\right)\psi dA + \int_V \boldsymbol{\phi}\frac{\partial}{\partial \mathbf{X}}\left(\frac{\partial W}{\partial C}\right) dV. \tag{26}$$

According to Clausius-Duhem inequality [17], free energy of a system that is thermodynamically consistent should always decrease or be equal to zero ($\delta\Lambda/\delta t \le 0$). This condition must apply to any random value of $\boldsymbol{\phi}$, ψ and $(\delta x/\delta t)$. Accordingly, all of the above terms should be less than or equal to zero. Short-range motion is much faster than long-range transport of the solvent. Relocations occurring locally in short-range motion are assumed to be instantaneous, hence first integrand in Eq. (26) should be zero. This gives

$$\mathbf{s}_{ij} = \frac{\partial W(\mathbf{F}, C)}{\partial \mathbf{F}_{ij}}. \tag{27}$$

When equilibrium in the elastomer is reached between fluid transport and mechanical loads, chemical potential becomes uniform. This can be achieved by equating the second integrand of Eq. (26) to zero, giving the following relation

$$\boldsymbol{\mu} = \frac{\partial W(\mathbf{F}, C)}{\partial C}. \tag{28}$$

When free energy density $W(\mathbf{F}, C)$ for a system is known, and assumption of local equilibrium is applied, (27) and (28) become the equations of state for the system. At equilibrium, stress ($\mathbf{s}_{ij}$) is a derivative of free-energy function with respect to deformation gradient, and chemical potential ($\boldsymbol{\mu}$) is defined as a derivative of free-energy with respect to concentration. Extended motion of polymer when volume change occurs due to fluid transport can be explained through kinetics, laying the foundation to develop an expression for flux due to gradient of chemical potential.

3.5 Kinematic constraint for network incompressibility

As explained earlier, elastomeric materials have larger values of bulk modulus as compared to shear modulus. They can therefore be considered as almost incompressible. Fluid transporting through the elastomer is assumed to be incompressible as well. Hence, overall response of swollen elastomer can be considered as incompressible. Jacobian can be defined as [6]:

$$J = \begin{bmatrix} \lambda_1 & & \\ & \lambda_2 & \\ & & \lambda_3 \end{bmatrix} \tag{29}$$

$$J \equiv \det(\mathbf{F}) = \lambda_1\lambda_2\lambda_3 = \frac{V_S}{V_o}.$$

Stretch is the ratio of initial and final linear dimensions that change due to deformation ($\lambda = l_f/l_0 = t_f/t_0 = h_f/h_0$). This definition in terms of stretches is based on invariants. Total volume is the sum of volume of dry polymer and volume of solvent molecules absorbed in the gel. Multiplying and dividing the resulting equation by number of solvent molecules, constraint for incompressibility can be determined as follows [18]:

$$\det(\mathbf{F}) = 1 + \frac{V_S}{V_o} = 1 + \frac{V_S}{N_S}\frac{N_S}{V_o},$$
$$\text{where } \det(\mathbf{F}) = 1 + \upsilon C. \tag{30}$$

In order to embed incompressibility, an integral should be added to the free-energy density function. Using Lagrange multiplier (p) to optimize the functions, applying the constraint [5] of Eq. (30), we get

$$\kappa = \int_V p\{1 + \upsilon C - \det(\mathbf{F})\}dV. \tag{31}$$

Differentiating Eq. (31) with respect to $\mathbf{F}$ and C, and using the identity $\partial\det(\mathbf{F})/\partial\mathbf{F} = \mathbf{F}^{-T}\det(\mathbf{F})$ [14], equations of state can be rewritten as follows:

$$\frac{\partial\kappa}{\partial\mathbf{F}} = -p\mathbf{F}^{-T}\det(\mathbf{F})$$
$$\frac{\partial\kappa}{\partial C} = p\upsilon. \tag{32}$$

Stress and chemical potential can now be defined as

$$\mathbf{s} = \frac{\partial W(\mathbf{F}, C)}{\partial\mathbf{F}} - p\mathbf{F}^{-T}\det(\mathbf{F})$$
$$\text{and} \quad \boldsymbol{\mu} = \frac{\partial W(\mathbf{F}, C)}{\partial C} + p\upsilon. \tag{33}$$

3.6 Kinetics

The only driving force causing swelling of the elastomer is the amount of fluid transport into the network, resulting in swelling and stretching of networks. This is the reason that kinetics needs to describe the fluid motion in terms of flux and gradient of chemical potential. Some authors consider gel as poro-elastic or porous and use Darcy's law to model the amount of fluid influx [10, 19]. In porous media, a body is considered to be made up of pores such that permeability is related to the square of pore size. But in a swollen elastomer, major portion of the body is comprised of solvent and when load is applied on it, maximum resistance comes from the solvent molecules as in diffusion. Kinetic theory has also been developed by Tanaka and Fillmore [20] by considering friction with rate-dependent swelling. This theory has limited applicability as it does not consider large deformations. Other authors define a kinetic law based on diffusion mechanism for solvent migrations in elastic material [4–6]. Mobility tensor should be dependent on deformation gradient and solvent concentration. For defining mobility tensor, we use the relations similar to Chester and Anand [10] and Hong et al. [5]:

$$\boldsymbol{\phi}_k = -\mathbf{M}_{KM}\frac{\partial\boldsymbol{\mu}}{\partial\mathbf{X}_M}. \tag{34}$$

Mobility tensor is positive definite and symmetric. Feynman et al. [21] derived an expression for diffusion in which flux is proportional to gradient of chemical potential. It explains the interrelationship of flux, fluid concentration, diffusion, and gradient of chemical potential in terms of true quantities:

$$\phi_i = -\frac{cD}{kT}\frac{\partial \boldsymbol{\mu}}{\partial \mathbf{x}_i}. \tag{35}$$

In engineering applications, a property based on its original or initial value is termed as nominal quantity. On the other hand, true quantities are based on instantaneous properties. For example, a structural member under the influence of load deforms, resulting in reduction of cross-sectional area. In nominal stresses, the cross-sectional area is assumed constant during the deformation and stress is determined, known as nominal or engineering stress. While in true stress, force is divided by the instantaneous area. True concentration and flux can be converted to nominal values through the relationships $c = C/\det(\mathbf{F})$ and $\phi = \mathbf{F}\boldsymbol{\phi}/\det(\mathbf{F})$, respectively. Also, using partial derivative chain rule, we can express gradient of chemical potential as

$$\frac{\partial \boldsymbol{\mu}}{\partial \mathbf{X}_M} = \frac{\partial \boldsymbol{\mu}}{\partial \mathbf{x}_i}\mathbf{F}_{iM}. \tag{36}$$

Replacing true concentration and true flux in Eq. (5) by their nominal values, we get

$$\frac{\mathbf{F}_{iK}}{\det(\mathbf{F})}\boldsymbol{\phi} = -\frac{C}{\det(\mathbf{F})}\frac{D}{kT}\frac{\partial \boldsymbol{\mu}}{\partial \mathbf{x}_i}. \tag{37}$$

Substituting Eqs. (34) and (36), along with the incompressibility condition from Eq. (30), into Eq. (37), relation for mobility tensor can be obtained:

$$\begin{aligned} \frac{\mathbf{F}_{iK}}{\det(\mathbf{F})}\mathbf{M}_{KM}\frac{\partial \boldsymbol{\mu}}{\partial \mathbf{X}_M} &= \frac{C}{\det(\mathbf{F})}\frac{D}{kT}\frac{\partial \boldsymbol{\mu}}{\partial \mathbf{x}_i} \\ \Rightarrow \mathbf{M}_{KM} &= \frac{D}{\upsilon kT}\mathbf{F}_{iK}^{-T}\mathbf{F}_{iM}^{-T}\{\det(\mathbf{F}) - 1\}. \end{aligned} \tag{38}$$

3.7 Thermodynamics of mixing

Thermodynamics involved in polymer and solvent mixing is very important in the development of an analytical model for polymer swelling. The theoretical basis for understanding the behavior of polymer solutions was established independently by Huggins [22] and Flory [23]. Flory-Huggins theory gives the energy of mixing for a pure polymer with a pure solvent in terms of enthalpy and entropy of mixing. **Figure 4** presents the lattice model for mixing of polymer and solvent.

Statistical explanation of entropy is used to determine the number of probable positions that the polymer can attain in the solution. Before mixing, both polymer and solvent have no unique state. After mixing, many probable states can be attained, given by

$$\boldsymbol{\Omega} = \frac{N!}{N_s!N_P!}. \tag{39}$$

Entropy for free energy of mixing can be specified according to Boltzmann expression as follows:

Figure 4.
Lattice model for mixing of polymer and solvent.

$$\Delta \mathbf{S} = k \ln \, \mathbf{\Omega}, \tag{40}$$

where k is the Boltzmann constant. Using logarithmic principles and Sterling's approximation, Eq. (40) can be rewritten in terms of volume fraction of polymer (Φ_P) and solvent (Φ_S) as follows:

$$\Delta \mathbf{S} = -k(N_s \ln \Phi_S + N_P \ln \Phi_P). \tag{41}$$

Enthalpy of mixing is defined in terms of a dimensionless entity known as polymer-solvent interaction parameter (χ). It measures the degree of interaction between polymer and solvent as well as polymer and polymer:

$$\Delta \mathbf{H} = \chi k T N_s \Phi_P. \tag{42}$$

For a closed system at constant pressure and temperature, Gibbs free energy can be defined as

$$\Delta \mathbf{G} = \Delta \mathbf{H} - T \Delta \mathbf{S}. \tag{43}$$

Putting Eqs. (41) and (42) in Eq. (43), we get the expression

$$\Delta \mathbf{G} = kT(N_s \ln \Phi_S + N_P \ln \Phi_P + \chi N_s \Phi_P). \tag{44}$$

Noting that ($\Phi_S + \Phi_P = 1$), and using Eq. (30), we get

$$\begin{aligned} \Phi_S &= 1 - \Phi_P = 1 - \frac{1}{J} \\ \Rightarrow \quad \Phi_S &= \frac{\upsilon C}{1 + \upsilon C}. \end{aligned} \tag{45}$$

Substituting Eqs. (45) into Eq. (44), and neglecting the middle term as it is very small compared to the remaining terms, the strain energy density function for mixing ($W_m = \Delta \mathbf{G}/V_o$) can be expressed as:

$$\begin{aligned} \mathrm{W}_m &= \frac{kT}{V_o}\left(N_s \ln \frac{\upsilon C}{1 + \upsilon C} + \chi N_s \frac{1}{1 + \upsilon C}\right) \\ \Rightarrow \quad \mathrm{W}_m &= \frac{kT}{\upsilon}\left(\upsilon C \, \ln \frac{\upsilon C}{1 + \upsilon C} + \chi \frac{\upsilon C}{1 + \upsilon C}\right). \end{aligned} \tag{46}$$

A similar expression has also been used by Chester and Anand [10] Duda et al. [24] and Kang and Huang [6]. Equation developed by Hong et al. [5] for strain energy density of mixing is different by only a constant value as compared to the above expression, which is negligible in deformation due to swelling. Flory and Rehner [25] postulated that for gels, free energy density is a combination of strain energy density function due to thermodynamics of mixing and stretching of polymer networks:

$$W(\mathbf{F}, C) = W_m(C) + W_S(\mathbf{F}). \tag{47}$$

Strain energy density function for stretching deformation can be based on either Gaussian or non-Gaussian theory. As described earlier, Gaussian theory mainly deals with the scenarios where short stretching of polymer chains is considered, so these models work well only for small deformations and are unable to match the deformation patterns at large strains. Most of the swelling models currently available use Gaussian statistics [8], and are therefore unable to give reasonable predictions for large swelling. Proposed analytical model is based on non-Gaussian theory, while Ogden strain energy function [9] is used to account for the limited extensibility. Ogden model is given by the following relation in terms of principal stretches:

$$W_S = \sum_{n=1}^{Q} \frac{\mu_n}{\alpha_n} \left(\lambda_1^{\alpha_n} + \lambda_2^{\alpha_n} + \lambda_3^{\alpha_n} - 3 \right). \tag{48}$$

Here, μ_n and α_n are material constants that are determined by fitting the experimental data, and Q is a positive definite integer. These material constants are related to shear modulus (G):

$$\sum_{n=1}^{Q} \mu_n \alpha_n = 2G. \tag{49}$$

For the proposed model, second degree Ogden strain energy function is used as it gives the closest prediction (as explained in Chapter 6). Expand Eq. (48) for second degree, and simplifying, we get the following relation:

$$\begin{aligned} W_S &= \frac{\mu_1}{\alpha_1} \left(\lambda_1^{\alpha_1} + \lambda_2^{\alpha_1} + \lambda_3^{\alpha_1} - 3 \right) + \frac{\mu_2}{\alpha_2} \left(\lambda_1^{\alpha_2} + \lambda_2^{\alpha_2} + \lambda_3^{\alpha_2} - 3 \right) \\ &\Rightarrow \mu_1\alpha_1 + \mu_2\alpha_2 = 2G. \end{aligned} \tag{50}$$

Putting Eqs. (46) and (50) into (47), we get the following strain energy density function:

$$\begin{aligned} W(\mathbf{F}, C) = &\frac{\mu_1}{\alpha_1} \left(\lambda_1^{\alpha_1} + \lambda_2^{\alpha_1} + \lambda_3^{\alpha_1} - 3 \right) + \frac{\mu_2}{\alpha_2} \left(\lambda_1^{\alpha_2} + \lambda_2^{\alpha_2} + \lambda_3^{\alpha_2} - 3 \right) \\ &+ \frac{kT}{\upsilon} \left(\upsilon C \ln \frac{\upsilon C}{1 + \upsilon C} + \chi \frac{\upsilon C}{1 + \upsilon C} \right) \end{aligned} \tag{51}$$

Knowing that deformation gradient can be represented as $\mathbf{F} = \lambda \mathbf{I}$ [13, 7], and differentiating Eq. (51) with respect to λ, we get

$$\frac{\partial W}{\partial \mathbf{F}} = \frac{\partial W}{\partial \lambda_i} = \mu_1 \lambda_i^{\alpha_1 - 1} + \mu_2 \lambda_i^{\alpha_2 - 1}. \tag{52}$$

Differentiation of Eq. (50) with respect to fluid concentration results in the following expression:

$$\frac{\partial W}{\partial C} = kT \left[\ln \frac{\upsilon C}{1 + \upsilon C} + \frac{1}{1 + \upsilon C} + \chi \frac{1}{(1 + \upsilon C)^2} \right]. \tag{53}$$

Substituting the expressions for $\partial W/\partial \mathbf{F}$ from Eq. (52), and $\partial W/\partial C$ from Eq. (53), into Eq. (33), and using (29), equations of state can be converted into the following form:

$$\mathbf{s}_i = \mu_1\lambda_i^{\alpha_1-1} + \mu_2\lambda_i^{\alpha_2-1} - p\lambda_i^{-1}\lambda_1\lambda_2\lambda_3$$
$$\text{and} \quad \boldsymbol{\mu} = kT\left[\ln\frac{\upsilon C}{1+\upsilon C} + \frac{1}{1+\upsilon C} + \chi\frac{1}{(1+\upsilon C)^2}\right] + p\upsilon. \tag{54}$$

Eq. (54) can be expanded for the three nominal stresses as follows:

$$\begin{aligned}\mathbf{s}_1 &= \mu_1\lambda_1^{\alpha_1-1} + \mu_2\lambda_1^{\alpha_2-1} - p\lambda_2\lambda_3\\ \mathbf{s}_2 &= \mu_1\lambda_2^{\alpha_1-1} + \mu_2\lambda_2^{\alpha_2-1} - p\lambda_1\lambda_3\\ \mathbf{s}_3 &= \mu_1\lambda_3^{\alpha_1-1} + \mu_2\lambda_3^{\alpha_2-1} - p\lambda_1\lambda_2.\end{aligned} \tag{55}$$

On the other hand, the two equations in (54) can be combined to obtain a constitutive relationship for modeling of swelling phenomenon:

$$\mathbf{s}_i = \mu_1\lambda_i^{\alpha_1-1} + \mu_2\lambda_i^{\alpha_2-1} - \frac{\lambda_i^{-1}\lambda_1\lambda_2\lambda_3}{\upsilon}\left[\boldsymbol{\mu} - kT\left\{\ln\frac{\upsilon C}{1+\upsilon C} + \frac{1}{1+\upsilon C} + \chi\frac{1}{(1+\upsilon C)^2}\right\}\right] \tag{56}$$

Eq. (56) provides a non-linear model for the phenomenon of swelling in elastomeric materials, considering non-Gaussian theory of polymer network stretching. This model takes mechanical as well as solvent properties as input, along with environmental conditions such as temperature, water salinity, swelling medium's coefficient of diffusion, polymer-solvent interaction parameter, etc.

In order to solve the equations of state, a case of free equilibrium swelling is considered. It is assumed that when dry elastomer is immersed in a solvent, it swells, and equilibrium is achieved after some time. At equilibrium between elastomer and diffusing solvent, chemical potential is negligible. Swelling is considered to be homogenous throughout the elastomer, making stretches equal in all directions. Replacing principal stretches with equivalent swelling stretch (λ_S), and setting up chemical potential equal to zero in Eq. (56), we get

$$s_i = \mu_1\lambda_S^{\alpha_1-1} + \mu_2\lambda_S^{\alpha_2-1} + \frac{\lambda_S^2}{\upsilon}kT\left[\ln\frac{\upsilon C}{1+\upsilon C} + \frac{1}{1+\upsilon C} + \chi\frac{1}{(1+\upsilon C)^2}\right]. \tag{57}$$

Using the identity of Eq. (30), we can write $1+\upsilon C = \lambda_S^3$ and $\upsilon C = \lambda_S^3 - 1$. Replacing in Eq. (57), and simplifying,

$$s_i = \mu_1\lambda_S^{\alpha_1-1} + \mu_2\lambda_S^{\alpha_2-1} + \frac{\lambda_S^2}{\upsilon}kT\left[\ln\left(1-\frac{1}{\lambda_S^3}\right) + \frac{1}{\lambda_S^3} + \chi\frac{1}{\lambda_S^6}\right]. \tag{58}$$

The elastomer is considered to be under no constraint. Application of this no-constraint condition on the elastomer results in no stress. Hence Eq. (58) can be transformed by embedding stress value equals to zero:

$$\mu_1\lambda_S^{\alpha_1-3} + \mu_2\lambda_S^{\alpha_2-3} + \frac{kT}{\upsilon}\left[\ln\left(1-\frac{1}{\lambda_S^3}\right) + \frac{1}{\lambda_S^3} + \chi\frac{1}{\lambda_S^6}\right] = 0. \tag{59}$$

4. Model validation

Validation of the model developed above requires the comparison of model predictions against experimental results. For this purpose, a series of experiments were performed under specific conditions. Some experimental results are used as input to the model (for evaluation of parameters), while other results are used for model validation. Logical flow of activities for model validation is shown in **Figure 5**. Swelling related experiments are discussed in detail in Chapters 3 and 7. Experimental work required to determine diffusion coefficients and the polymer-solvent interaction parameter and is explained below.

4.1 Experimental investigation

Experiments were conducted (already described in Chapter 7) on disc samples of two different water swelling elastomers, in salt solutions of low and high salinities (0.6% and 12%), at room temperature and 50°C. During the one-month swelling period, readings (volume, thickness, mass, and hardness) were taken before swelling and after 1, 2, 4, 7, 10, 16, 23, and 30 days of swelling. Stres-strain relations from compression and bulk tests were used to determine values of bulk modulus, and different structural properties. Evaluation of other solvent and polymer properties (diffusion coefficient, polymer-solvent interaction parameter, and molar volume of swelling solvent) is discussed below.

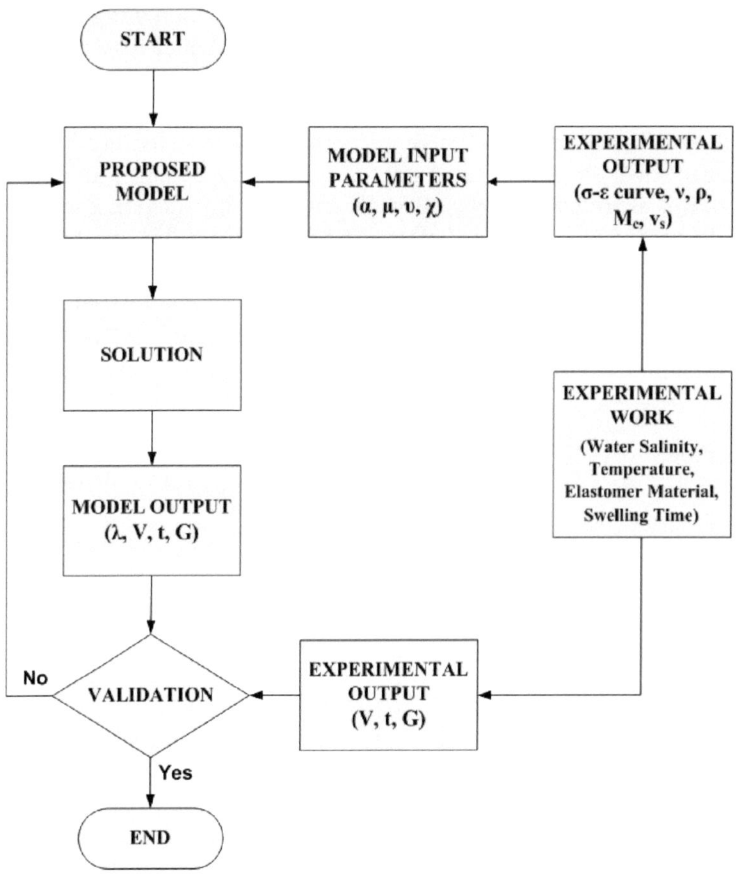

Figure 5.
Flow diagram of activities required for model validation.

4.1.1 Diffusion coefficient

When a dry polymer is immersed into water of a specific salinity, solvent starts diffusing into the polymer due to chemical potential gradient. Diffusion coefficients for low and high salinity solutions should be known in order to correctly predict the amount of swelling. Stokes-Einstein formula [26] is used here to evaluate the diffusion coefficients for low and high salinity solutions.

For a fluid with no flow separation, the drag force is given by

$$F_D = \frac{24\left(\frac{\rho v^2}{2}\right)A}{Re}. \tag{60}$$

We know that Reynold's number and cross-sectional area are given by $Re = Dv\rho/\eta$ and $A = \pi D^2/4$ respectively. Therefore,

$$F_D = 6\pi R\eta v. \tag{61}$$

It is known that drag force can also be represented as

$$F_D = \zeta v, \tag{62}$$

where ζ is Stoke's friction factor, and can be written as

$$\zeta = 6\pi R\eta. \tag{63}$$

The diffusion coefficient (D) is given by Einstein's equation as

$$D = \frac{kT}{\zeta}. \tag{64}$$

Using Stoke's friction factor, this becomes

$$D = \frac{kT}{6\pi R\eta}. \tag{65}$$

Here, k is the Boltzmann constant, T is the absolute temperature, R is the radius of solvent particle, and η is the viscosity.

In order to determine diffusion coefficient, viscosity and density of water at low and high salinities are required for all test conditions. To determine the viscosity, Cannon-Fenske [27] apparatus (**Figure 6**) of size 50 is used. Liquid (whose viscosity is to be determined) is filled in the apparatus slightly above the top-bulb. Fluid is then allowed to flow downwards, pass through the tube, and collect at the bigger bulb at the bottom. Time is recorded for fluid to cross two marks at the top and bottom of the second bulb. Readings are taken at the same temperature at which all swelling related experiments are conducted.

Density of elastomer samples is calculated by determining their mass and volume after each swelling period, for both salinities and at both temperatures. Mass of the elastomer is increased by inflow of salt water into the material. At the same time, volume is increased through swelling. As both mass and volume increase almost proportionally, density does not change too much. Calculated density values are 0.967 g/cm^3 and 1.016 g/cm^3 for low salinity water at room temperature and 50°C, and 1.05 g/cm^3 and 1.0898 g/cm^3 for high salinity water at the two temperatures respectively.

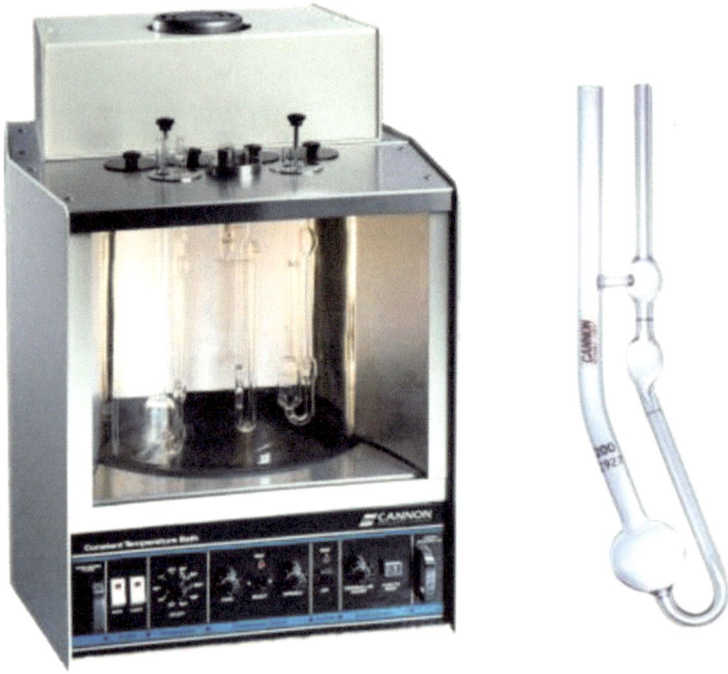

Figure 6.
Cannon-Fenske apparatus used for viscosity measurement.

Mean time for low and high salinity water to pass from the first to the second mark in Cannon-Fenske apparatus was recorded as 157 and 170 sec for 50°C, and 219 and 260 sec at room temperature, respectively. Multiplying by the apparatus factor of 0.004, we get viscosity (centistokes) values of 0.628 and 0.68 at 50°C, and 0.876 and 0.1.04 at room temperature, for low and high salinities respectively. Multiplying this dynamic viscosity with density yields the kinematic viscosity in Pascal-second units. Substituting these values into Eq. (65) gives the required value of diffusion coefficient for each condition. These experimentally determined values of viscosity and diffusion coefficient are summarized in **Table 1**.

It can be seen that the diffusion coefficient has higher value in low salinity water as compared to high salinity water, for both room and 50°C temperatures. This higher diffusion amount causes faster swelling rate when the elastomer is kept in low salinity solutions. For same salinity, diffusion coefficient has higher values for

Viscosity (Pa-s x10^{-3})		**Temperature**	
		Low (room)	**High (50 °C)**
Salinity	Low (0.6%)	0.89	0.607
	High (12%)	1.1334	0.714
Diffusion coefficient(m^2/s x10^{-10})		**Temperature**	
		Low (room)	**High (50 °C)**
Salinity	Low (0.6%)	8.22	11.65
	High (12%)	6.46	9.9

Table 1.
Viscosity and diffusion coefficient at two salinities and two temperatures.

higher temperatures, again matching the observed trend for higher amount of elastomer swelling at higher temperature.

Though the mechanism of swelling has been discussed earlier (Chapters 3 and 7), let us revisit the issue in the context of diffusion. When sodium chloride dissolves in water, crystalline structure of sodium and chlorine transform into positive and negative ions surrounded by water molecules. Water molecules close to the ions have strong attraction. When polymer is exposed to brine solution, water molecules diffuse into the empty spaces and begin to fill the voids, resulting in swelling. Ions surrounded by water molecules also enter the polymer but somewhat slowly due to stronger attraction. Ultimately, the polymer is filled by water molecules and ions. Low salinity solution has less number of ions, hence water fills the spaces more quickly as compared to high salinity solution. That is why diffusion takes place at a slower rate in high salinity salt solution.

4.1.2 Polymer-solvent interaction parameter

Polymer-solvent interaction parameter (χ), also known as Flory-Huggins interaction parameter, is an important dimensionless temperature-based property of a polymer which controls the amount of swelling. If χ increases, liquid seepage takes place leading to de-swelling or contraction of the elastomer. A decrease in χ-value leads to swelling of the elastomer (volume increase). Different experimental studies are available in published literature which discuss methods of determining χ. Orwoll and Arnold [28] use inverse gas chromatography. Papageorgiou et al. [29] apply differential scanning calorimetry. Silva et al. [30] use measurement of melting temperature of the polymer blend. Clarke et al. [31] use micelle spacings with secondary ion mass spectroscopy, contact angle of blend droplet, and neutron reflectometry. In all experimental methods, the polymer needs to dissociate in order to determine χ for a particular polymer-solvent mixture. Both elastomer materials used in swelling experiments were found to be insoluble in all solvents available (polar as well non-polar solvents). An in-depth search was carried out to find a method for determination of χ that does not require elastomer dissociation. Treloar [8] gives a relationship between molecular mass and volume swelling ratio, density of polymer, molar volume of swelling liquid, and polymer-solvent interaction parameter:

$$\begin{gathered}\frac{\rho V_1}{M_c}\left(v_s{}^{1/3}-\frac{\nu}{2}\right)+\left[\ln(1-v_s)+v_s+\chi v_s{}^2\right]=0\\ \Rightarrow\ \chi=-\frac{1}{v_s{}^2}\left[\frac{\rho V_1}{M_c}\left(v_s{}^{1/3}-\frac{v_s}{2}\right)+\ln(1-v_s)+v_s\right].\end{gathered}\tag{66}$$

The above Eq. (66) is used in this work to determine the value of χ for each data set. Molecular mass (M_c) is already determined using experiments described in Chapter 7. This unique method for the determination of the interaction parameter (χ), using mechanical and structural properties of a polymer, has not been used in any of the published works. Molar volume (V_1) of water is 0.00018 m^3. Values of density of elastomer after different swelling periods are given above. **Table 2** lists the experimentally determined values of volume swelling ratio (v_s), which is the ratio of initial to swelled volume.

Figure 7 shows the variation of χ against swelling time (days) for both materials in low and high salinity brine. For all the cases, χ values tend to decrease with swelling time, following the default trend [13]. It can also be seen that χ values drop more in low salinity brine, in line with the earlier observation that these elastomers swell more in brines of lower concentration. Initially faster and then more gradual

Days of Swelling	Volume Swelling Ratio (v_s)			
	Material-A		Material-B	
	Low-Salinity	High-Salinity	Low-Salinity	High-Salinity
1	0.455	0.543	0.494	0.646
3	0.392	0.505	0.403	0.525
6	0.323	0.465	0.318	0.474
9	0.291	0.431	0.292	0.442
15	0.245	0.365	0.264	0.438
22	0.234	0.347	0.249	0.389
30	0.234	0.338	0.232	0.376

Table 2.
Volume swelling ratio for materials A *and* B.

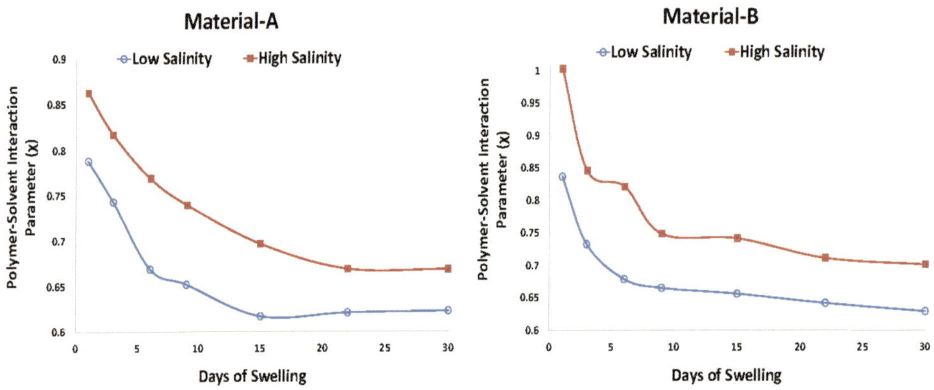

Figure 7.
Variation of polymer-solvent interaction parameter (χ) against swelling time; materials A *and* B*; low and high salinity.*

decrease in χ also matches the behavior of these quick-swelling elastomers. Slight fluctuation of χ variation for material-*B* is also consistent with earlier observations about swelling behavior.

4.2 Computational code

For the solution of the final Eq. (59) of the developed model, and to extract different values required for analysis, a MATLAB code is written. This requires inputs such as material coefficients, temperature, volume per solvent molecule, interaction parameter, etc. Logical sequence of different steps involved are shown in **Figure 8**. Using all the input values, and going through the various steps of the iterative scheme, the code estimates the magnitudes of swelling stretches after each swelling period (days) and for each material and each salinity (low and high). Comparison of different outcomes of the model with experimental results is discussed in next section.

5. Analysis of results

New model starts with Ogden-2 as its basis, but is changed into a totally new relationship (Eq. 59) by the introduction of terms for diffusion and

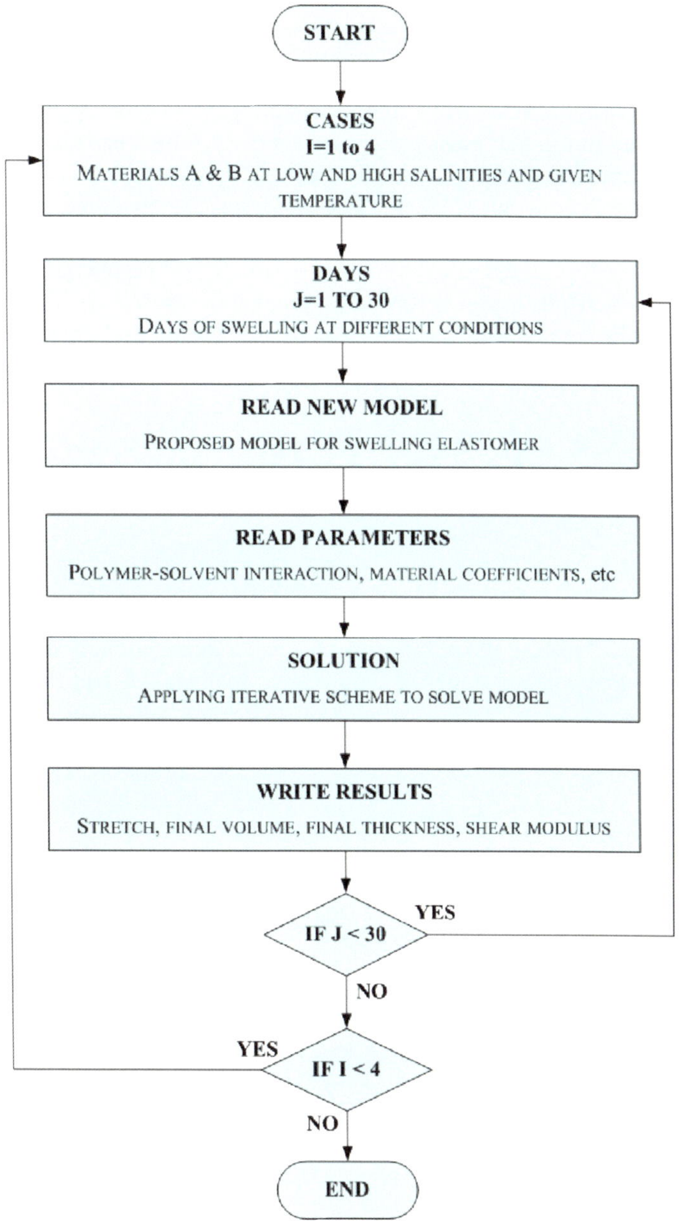

Figure 8.
Logical sequence of steps involved in the MATLAB code.

thermodynamics of mixing. Material coefficients for Ogden model (μ_1, μ_{12}, α_1, α_2) are determined from experimental stress–strain curves and Poisson's ratio (υ). Polymer-solvent interaction parameter (χ) is determined from mechanical, structural, and chemical properties: density (ρ) and chain molecular mass (M_c) of the elastomer, molar volume of water (V_1), and volume swelling ratio (v_s). Other input parameters needed are temperature (T), Boltzmann constant (k), and swelled volume per solvent molecule (υ). Through an iterative solution, major output from the new model is the value of swelling stretch (λ) at each stage of swelling, from which the amount of volume and thickness change and the shear modulus (G) are calculated.

5.1 Elastomer stretch

Figure 9 gives the variation of stretch for materials *A* and *B* under both salinities. Results show an increase in stretch throughout the swelling period for both the materials, sharply in the beginning, then more slowly. This result is in good agreement with the fact that swelling leads to volume increase, which is represented by an increase in the stretch value. Higher stretch values are observed in low salinity solvent, in line with the actual behavior of swelling elastomers which swell more in low salinity due to higher chemical potential gradient and higher diffusion coefficient. Minor fluctuations in the variation of amount of stretch are also similar to the fluctuations in volume and thickness swelling reported and analyzed in Chapters 3 and 7.

5.2 Volume swelling

Volume after swelling can be calculated from the amount of stretch determined from the model:

$$V_{Model} = {\lambda_S}^3 * V_o. \quad (67)$$

Figures 10 and **11** show the comparison between experimental and model-predicted volume at various stages of swelling for materials *A* and *B* under the two

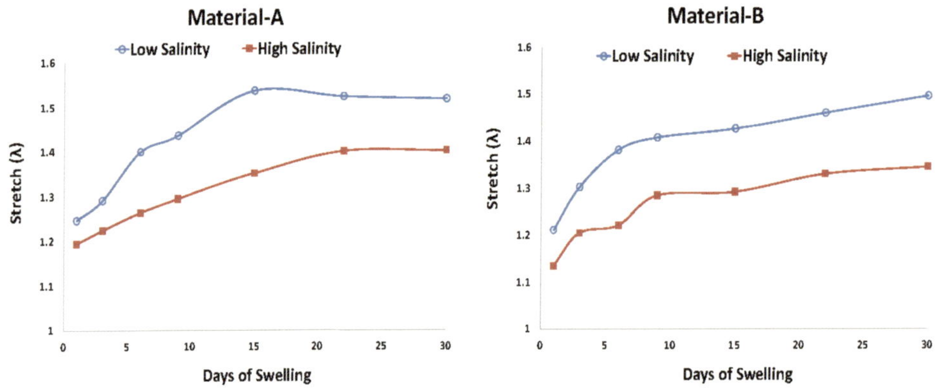

Figure 9.
Amount of stretch vs. swelling time for materials A *and* B*; both salinities.*

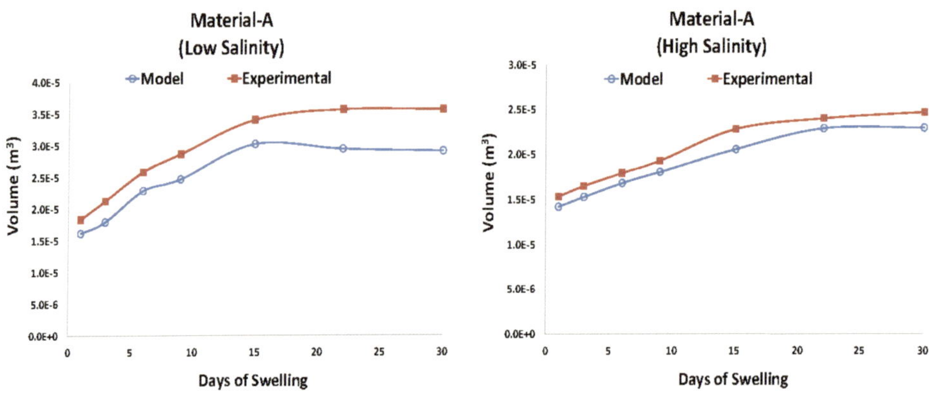

Figure 10.
*Model-predicted and experimental results for volume swelling; material-*A*; both salinities.*

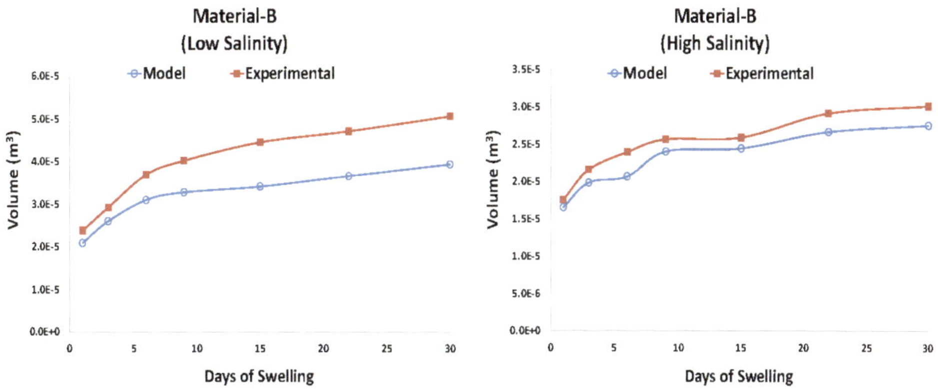

Figure 11.
*Model-predicted and experimental results for volume swelling; material-*B*; both salinities.*

salinities. Following the pattern of experimental results, predicted amount of elastomer volume increases with swelling time (days). Slight fluctuations in volume swelling are also in line with experimental results, and due to reasons explained earlier. There is a small difference between experimental and model results, the model consistently underestimating the value of volume swelling. This underestimation gives a more conservative estimate of swelling amount, which is better and safer in terms of seal design and development.

5.3 Thickness swelling

Final thickness after each swelling stage can be calculated using swelling stretch as follows:

$$\mathrm{t}_{Model} = \lambda_S * t_o. \tag{68}$$

Comparison between experimental and predicted values of thickness at various stages of swelling is shown in **Figures 12** and **13** for materials *A* and *B* in both salinities. There is a gradual increase in the amount of thickness with more days of swelling, with a reasonably good agreement between predicted values and experimental results. As observed earlier, there is more thickness swelling under lower salinity brine.

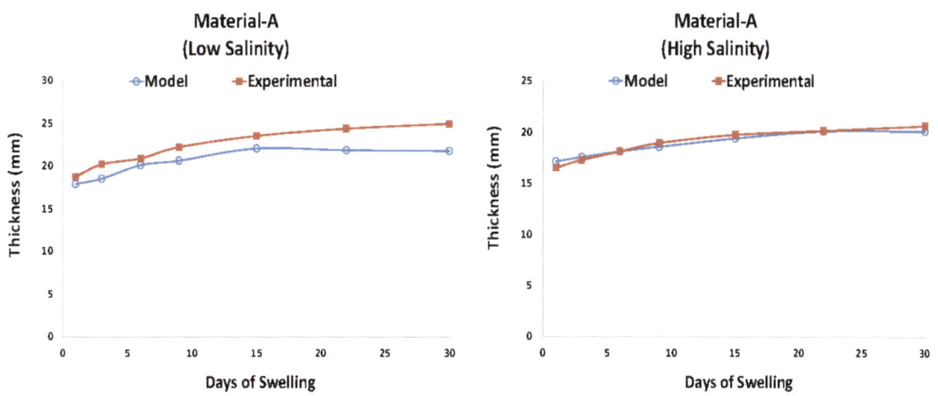

Figure 12.
*Model-predicted and experimental results for thickness swelling; material-*A*; both salinities.*

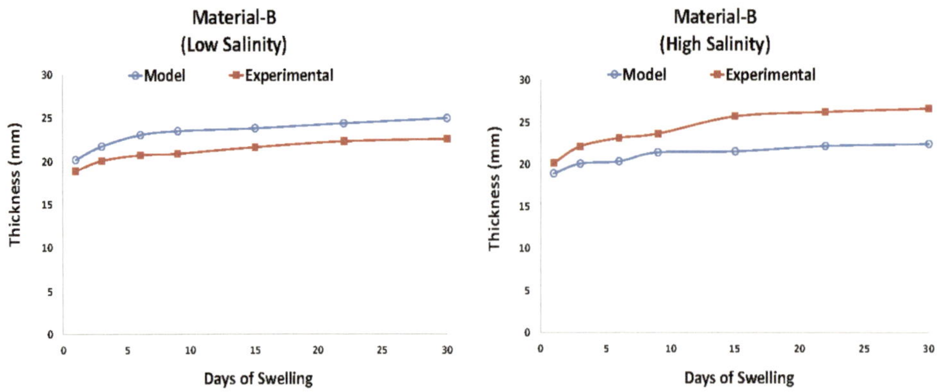

Figure 13.
*Model-predicted and experimental results for thickness swelling; material-*B*; both salinities.*

5.4 Shear modulus

Amount of stretch predicted from the model can be used to determine the value of shear modulus using the following relationship:

$$G_{Model} = -\frac{N_c \upsilon\left(\mu_1 \lambda_S^{\alpha_1 - 3} + \mu_2 \lambda_S^{\alpha_2 - 3}\right)}{\left[\ln\left(1 - \frac{1}{\lambda_S^3}\right) + \frac{1}{\lambda_S^3} + \chi \frac{1}{\lambda_S^6}\right]}. \tag{69}$$

Predicted and experimental values of shear modulus are plotted against swelling time in **Figures 14** and **15** for the two materials and salinities. For further comparison, variation of shear modulus through numerical simulation (Chapter 7) using the best available hyperelastic material model (Ogden-2) is also shown. ***It is reassuring to observe that there is a very good agreement between values predicted by the new model and experimental results, even in portions where there are major fluctuations. Also, predictions from the new model are much closer than numerical simulations using the best hyperelastic material model.*** Even though hyperelastic material models are based on the theory of shear, simulated values of *G* using the current best model (Ogden-2) show notable variations from the experimental values. As discussed at the end of Chapter 6, hyperelastic models can be used to predict the behavior of swelling elastomers only as a last resort, with some errors.

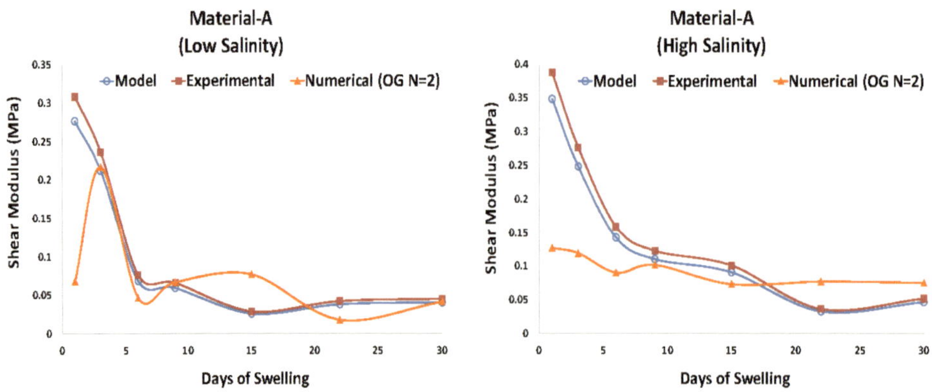

Figure 14.
*Variation of shear modulus against swelling time from new model, experiments, and FE simulation (Ogden-2); material-*A*; both salinities.*

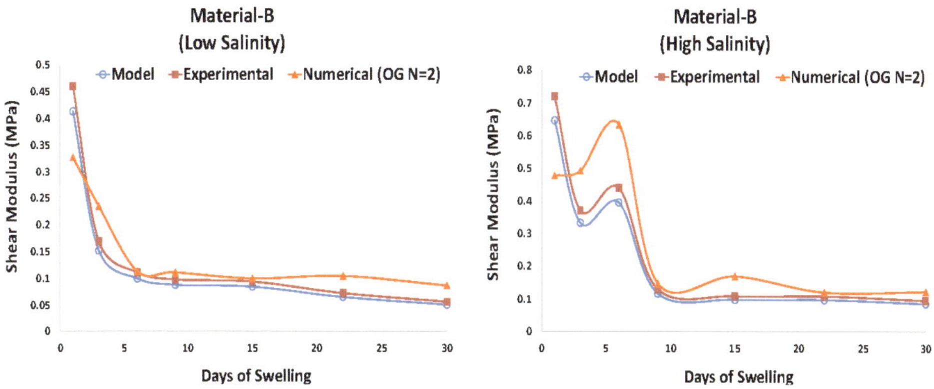

Figure 15.
Variation of shear modulus against swelling time from new model, experiments, and FE simulation (Ogden-2); material-B; both salinities.

This is mainly because these models do not include the effect of diffusion and thermodynamics of mixing. As the new model includes all these effects, its predictions are much closer to the actual values.

This new model is not material-specific, and can be applied to any situation where swelling is taking place. As the use of swelling elastomers in the petroleum industry is a major category of such applications, these results are used here to validate the model predictions. However, the same model can be used to predict the behavior of other soft materials under swelling such as tissue, cartilage, and other biological materials.

6. Conclusions

A new analytical model has been developed for predicting the behavior of swelling elastomers, based on nonlinear and non-Gaussian continuum mechanics, different balance laws for forces and solvent (including diffusion), and the thermodynamics of mixing. Including energy, diffusion, and hyperelastic terms, this new model can be used for both constrained and free swelling. Boundary conditions for free swelling are incorporated into the model. A MATLAB code is then developed for model solution. New experiments have been performed to determine input values such as viscosity of swelling medium and polymer-solvent interaction parameter. Stretch values predicted by the model are used to determine volume and thickness swelling, and variation of shear modulus. Model predictions have good agreement with experimental results, much closer than numerical simulation based on best existing hyperelastic material model.

Author details

Sayyad Zahid Qamar[1]*, Maaz Akhtar[2] and Tasneem Pervez[1]

1 Mechanical and Industrial Engineering Department, Sultan Qaboos University, Muscat, Oman

2 Mechanical Engineering Department, N.E.D. University of Engineering and Technology, Karachi, Pakistan

*Address all correspondence to: sayyad@squ.edu.om

References

[1] Akhtar M, Qamar SZ, Pervez T, Al-Jahwari FK (2018) "Performance Analysis of Swelling Elastomer Seals," Petroleum Science and Engineering, 165 (2018), p 127–135

[2] Okumura D, Chester SA (2018) "Ultimate swelling described by limiting chain extensibility of swollen elastomers," International Journal of Mechanical Sciences, 144, p 531–539

[3] Landsgesell J, Sean D, Kreissl P, Szuttor K, Holm C (2019) "Modeling Gel Swelling Equilibrium in the Mean Field: From Explicit to Poisson-Boltzmann Models," Physical Review Letters, 122, 208002

[4] Hong W, Liu Z, Suo Z (2009) "Inhomogeneous Swelling of a Gel in Equilibrium with a Solvent and Mechanical Load, International Journal of Solids and Structures, 46(17), 3282–3289

[5] Hong W, Zhao X, Zhou J, Suo Z (2008) "A Theory of Coupled Diffusion and Large Deformation in Polymeric Gels," Journal of the Mechanics and Physics of Solids, 56(5), 1779–1793

[6] Kang MK, Huang R (2010) "A Variational Approach and Finite Element Implementation for Swelling of Polymeric Hydrogels under Geometric Constraints," Journal of Applied Mechanics, 77(6), 061004

[7] Lucantonio A, Nardinocchi P, Teresi L (2013) "Transient Analysis of Swelling-Induced Large Deformations in Polymer Gels," Journal of the Mechanics and Physics of Solids, 61(1), 205–218

[8] Treloar LRG (1975) The Physics of Rubber Elasticity, Oxford University Press

[9] Ogden R (1972) "Large Deformation Isotropic Elasticity — On the Correlation of Theory and Experiment for Incompressible Rubberlike Solids," Proceedings of the Royal Society of London. A. Mathematical and Physical Sciences, 326(1567), 565–584

[10] Chester SA, Anand L (2010) "A Coupled Theory of Fluid Permeation and Large Deformations for Elastomeric Materials," Journal of the Mechanics and Physics of Solids, 58(11), 1879–1906

[11] Flory PJ (1953) Principles of Polymer Chemistry, Cornell University Press

[12] Crank J (1979) The Mathematics of Diffusion, Oxford University Press

[13] Doi M (2009) "Gel Dynamics," Journal of the Physical Society of Japan, 78(5)

[14] Mase GT, Mase GE (2010) Continuum Mechanics for Engineers, CRC Press

[15] Bucalem ML, Bathe K-J (2011) The Mechanics of Solids and Structures: Hierarchical Modeling and the Finite Element Solution, Springer

[16] Gurtin ME, Fried E, Anand L (2010) The Mechanics and Thermodynamics of Continua, Cambridge University Press

[17] Frémond M (2006) "The Clausius-Duhem Inequality, an Interesting and Productive Inequality," in: Alart P, Maisonneuve O, Rockafellar RT (eds) Nonsmooth Mechanics and Analysis. Advances in Mechanics and Mathematics, vol 12. Springer, Boston

[18] Drozdov A, Christiansen J (2013) "Constitutive Equations in Finite Elasticity of Swollen Elastomers," International Journal of Solids and Structures, 50(9), 1494–1504

[19] Biot MA (1941) "General Theory of Three-Dimensional Consolidation," Journal of Applied Physics, 12, 155–164

[20] Tanaka T, Fillmore DJ (1979) "Kinetics of Swelling Gels," Journal of Chemical Physics, 70(3), 1214

[21] Feynman R, Leighton R, Sands M (1963) The Feynman Lectures on Physics, Addison-Wesley

[22] Huggins ML (1942) "Some Properties of Solutions of Long-Chain Compounds," The Journal of Physical Chemistry, 46(1), 151–158

[23] Flory PJ (1942) "Thermodynamics of High Polymer Solutions," Journal of Chemical Physics, 10, 51–61

[24] Duda FP, Souza AC, Fried E (2010) "A Theory for Species Migration in a Finitely Strained Solid with Application to Polymer Network Swelling," Journal of the Mechanics and Physics of Solids, 58(4), 515–529

[25] Flory P, Rehner J (1943) "Statistical Mechanics of Cross-Linked Polymer Networks: I. Swelling," Journal of Chemical Physics, 11, 521–526

[26] Einstein A (1905) "On the Movement of Small Particles Suspended in Stationary Liquids Required by the Molecular-Kinetic Theory of Heat," Annalen der Physik, 17 (1905), 549–560

[27] Cannon M, Fenske M (1938) "Viscosity Measurement," Industrial & Engineering Chemistry Analytical Edition, 10(6), 297–301

[28] Orwoll RA, Arnold PA (2007) "Polymer–Solvent Interaction Parameter χ," in JE Mark (ed) Physical Properties of Polymers Handbook, p 233–257, Springer, New York

[29] Papageorgiou GZ, Panayiotou C (2011) "Crystallization and Melting of Biodegradable Poly(propylene suberate)," Thermochimica Acta, 523 (1–02), p 187–199

[30] Silva L, Tognana S, Salgueiro W (2013) "Miscibility in Crystalline/Amorphous Blends of Poly(3-hydroxybutyrate)/DGEBA," Journal of Polymer Science, Part B: Polymer Physics, 51, 680–686

[31] Clarke CJ, Eisenberg A, La Scala J, Rafailovich MH, Sokolov J, Li Z, Qu S, Nguyen D, Schwarz SA, Strzhemechny Y, Sauer BB (1997) "Measurements of the Flory–Huggins Interaction Parameter for Polystyrene–Poly(4-vinylpyridine) Blends," Macromolecules, 30, 14, 4184–4188